# Death's Twilight Kingdom

## Volume 2

## The Secret World of U.S. ICBM's, Re-entry Vehicles, and Dynamic Strategic Nuclear Forces Control

"We ate the food,
We drank the wine,
Everyone having a real good time.
Except You.
You were talking about the End of the World."

-- "Until the End of the World" (1991)
U2

by Yogi Shan

**2<sup>nd</sup> Edition**

**Version 2.35**

**ISBN-13: (Printed Paperback)**
**ISBN-10: (Kindle device, e-version "Soft Copy")**

**Printed by: CreateSpace.com**
**Distributed by: CreateSpace.com, amazon.com, amazon.ca, amazon.co.uk, and many others**

**Comments, inquiries, insults, and ordering info for this book to: yogishan2000@yahoo.ca**

Dedicated to my children, Oz and "A.C.", who will perhaps now understand that their father was not some crazed, wild-eyed, gun-nut who loved caressing his AK-47, and who babbled endlessly about freedom, corruption, revolution, and other strange things during his frequent rants.

To my children: may you live, learn, and love with less pain than I did.

You are the only two things of eternal value that I ever produced. And it was only by accident – it was only their mothers that saw the value in me that I was too beaten down to even dream of.

**Other CreateSpace books by Yogi Shan:**

"Death's Twilight Kingdom: Volume 1 – The Secret World of U.S. Nuclear Weapon "Design Data""

"A Stranger in Strange Times: A Story of One Man's Unconventional Philosophy"

"Urban Guerrilla Bedtime Stories: The Inside Story of Modern Arms Design"

"Declassified Nerve Gas Production Processes"

"The Corrected and Updated 'Anarchist Cookbook' "

**Upcoming book by Yogi Shan:**

"LSD, 'Speed', and Synthetic Heroin: A Popular History, Written by a Chemist"

# Table of Contents

# Figures

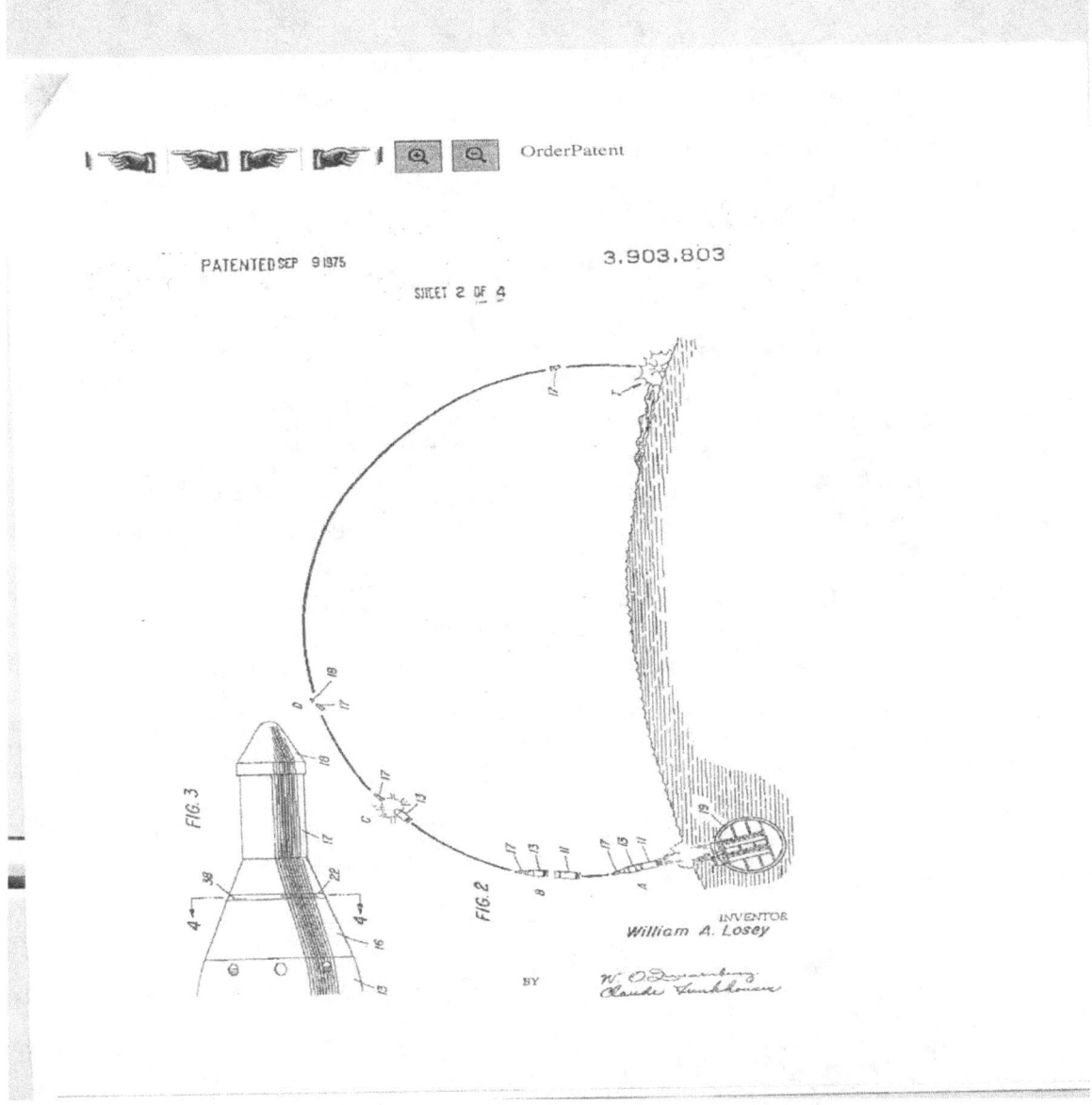

FIG. 3

FIG. 2

INVENTOR
William A. Losey

BY

Figure A:  Polaris SLBM Launch-to-Target Sequence

United States Patent [19]

Hall et al.

[11] Patent Number: 4,470,562

[45] Date of Patent: Sep. 11, 1984

[54] POLARIS GUIDANCE SYSTEM

[75] Inventors: Eldon C. Hall, Wollaston; Joseph D. Sabo, Arlington; Samuel A. Forter, Hingham; Ralph R. Ragan, Lincoln; J. H. Laning, West Newton; David G. Hoag, Medway; Wallace E. Vander Velde, Winchester; Daniel J. Lickly, Melrose; Edward M. Copps, Jr., Arlington, all of Mass.

[73] Assignee: The United States of America as represented by the Secretary of the Navy, Washington, D.C.

[21] Appl. No.: 502,717

[22] Filed: Oct. 22, 1965

[51] Int. Cl.³ ........................ F41G 7/36; F42B 15/18
[52] U.S. Cl. .................................. 244/3.2; 244/3.22
[58] Field of Search ................................. 35/150.25;
244/3.2–3.22, 77; 60/230

[56] References Cited

U.S. PATENT DOCUMENTS

| 2,946,539 | 7/1960 | Fischel | 244/3.2 |
| 3,003,312 | 10/1961 | Jewell | 60/230 |
| 3,078,042 | 2/1963 | Grado | 244/3.2 |
| 3,164,340 | 1/1965 | Slater et al. | 244/3.2 |
| 3,219,293 | 11/1965 | Parker et al. | 244/3.2 |
| 3,231,726 | 1/1966 | Williamson | 244/3.2 |
| 3,249,324 | 5/1966 | Coffman | 244/3.2 |
| 3,301,508 | 1/1967 | Yamron | 244/3.2 |

Primary Examiner—Charles T. Jordan
Attorney, Agent, or Firm—Robert F. Beers; Arthur L. Branning

[57]        ABSTRACT

An inertial guidance system for a rocket powered ballistic missile which provides navigation and control while the missile is proceeding in its flight trajectory. Signals are generated within the confines of the missile and without outside information representative of angular and linear motion of the missile with respect to an inertial frame of reference, a digital computer compares these signals against a preset parameter stored in the computer to provide control signals along the pitch and yaw axes. An autopilot uses these control signals to provide mechanical control movements, and an exhaust nozzle deflection arrangement connected to the autopilot deflects the exhaust of the rocket for guidance along a predetermined flight trajectory.

31 Claims, 31 Drawing Figures

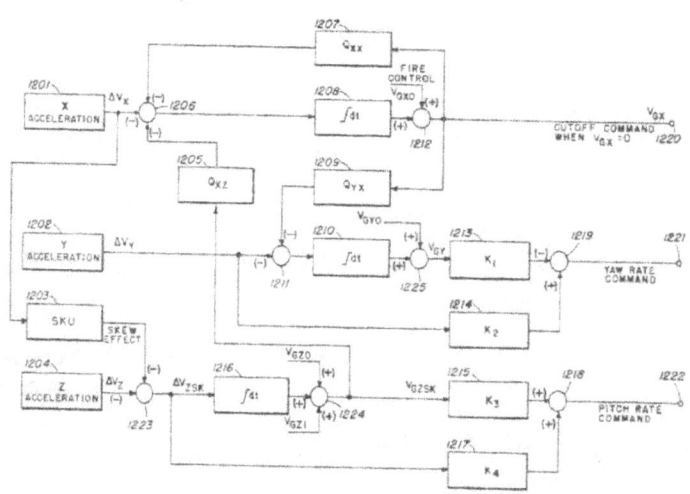

Figure B:  Polaris Guidance System Computer Patent, Front Page

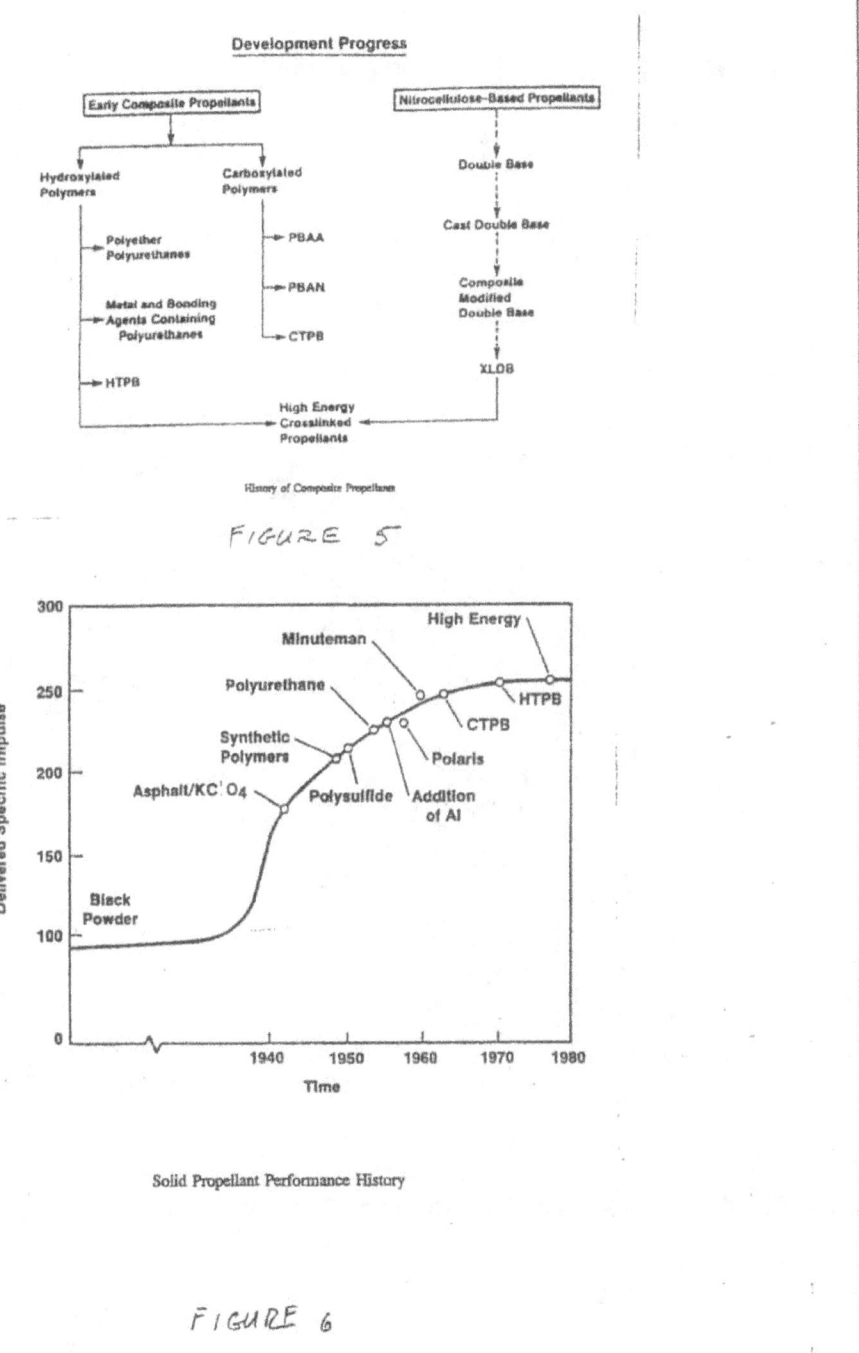

Figure C: Evolution of ICBM/SLBM Solid Propellants

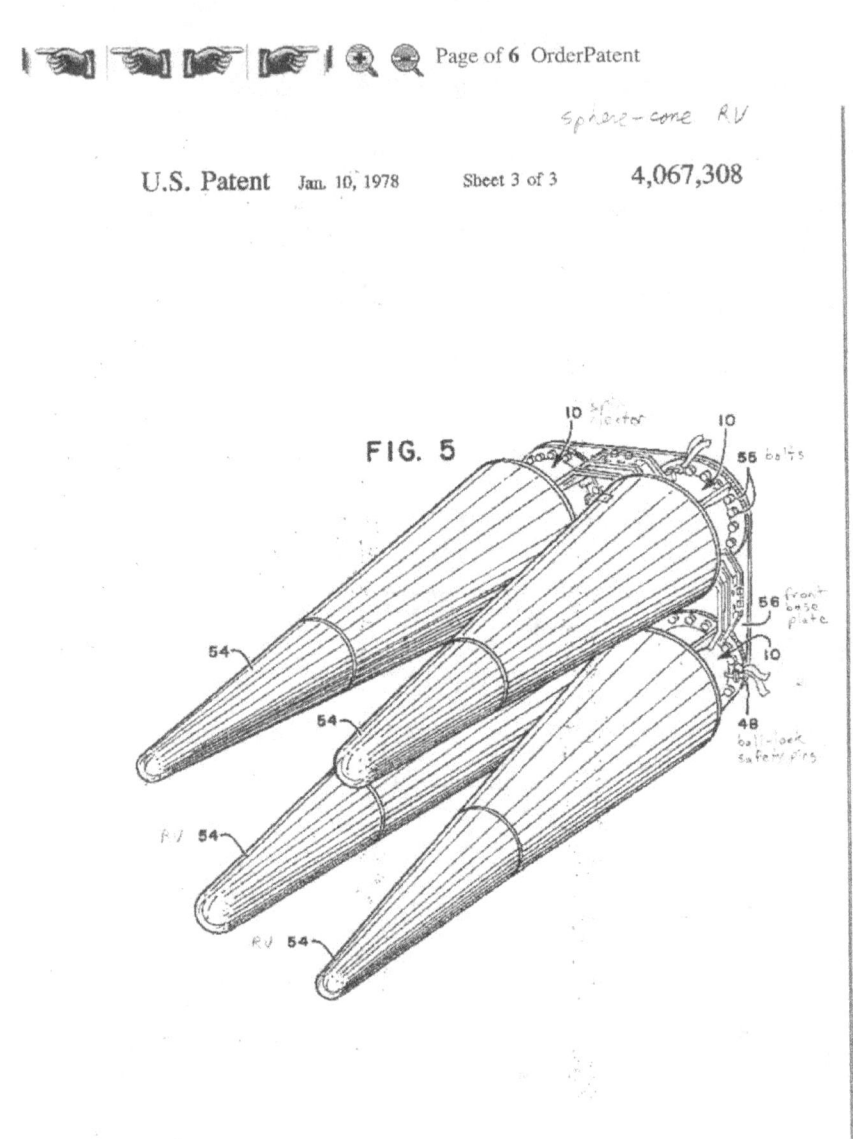

FIG. 5

Figure D: U.S. Multiple Independently-targeted Reentry Vehicle (MIRV) Load

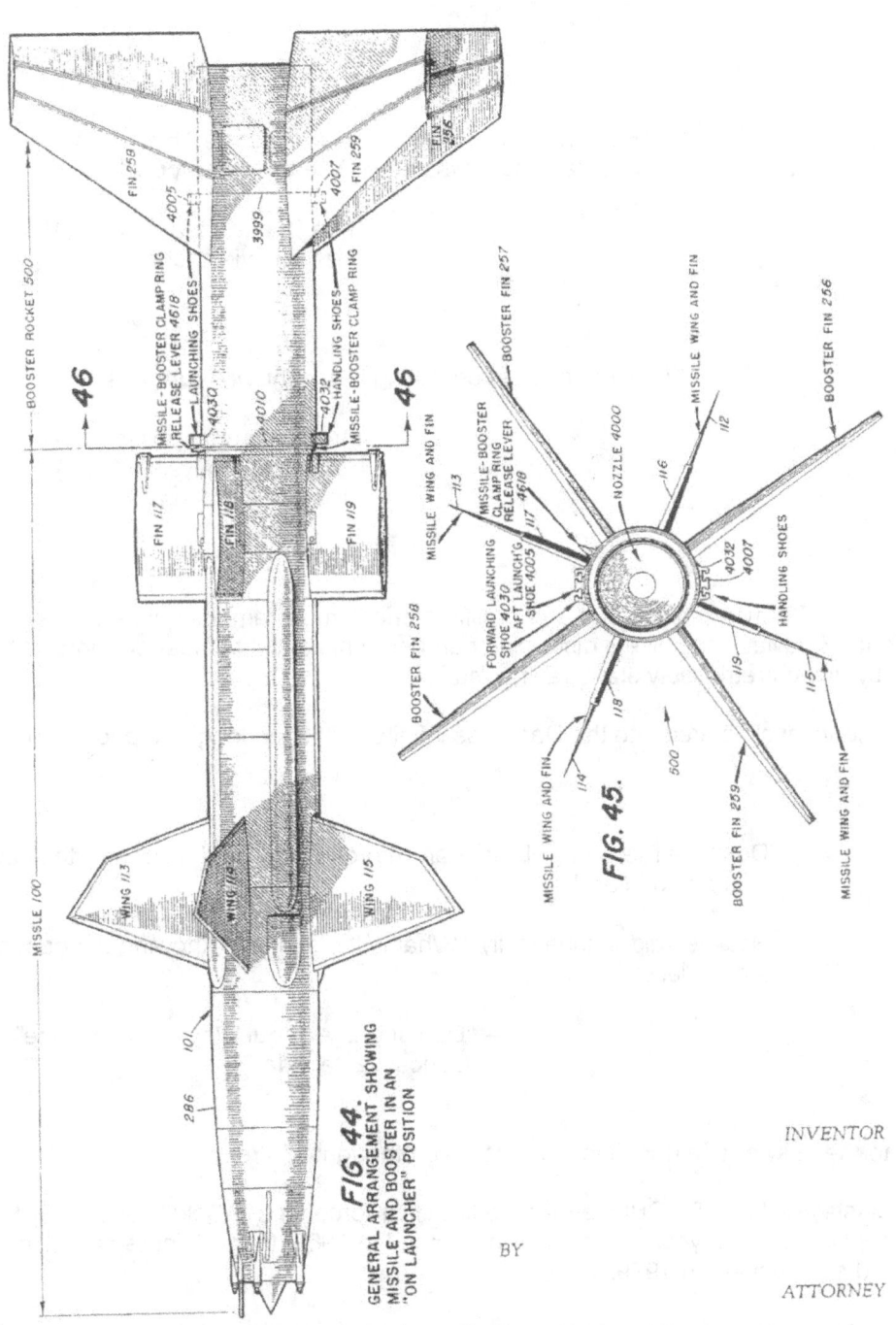

FIG. 44.
GENERAL ARRANGEMENT SHOWING MISSILE AND BOOSTER IN AN "ON LAUNCHER" POSITION

FIG. 45.

INVENTOR

BY

ATTORNEY

:Figure E:  TALOS, Mark 30-Armed, Air-Breathing Atomic Missile Patent, Page 1

# Perfecting Armageddon

## "Reconfiguration": The Dynamic Control of U.S. Land-Based ICBM Forces

"Be careful how long you stare into the Abyss. For one day, the Abyss may stare back into **you**."

-- "Thus Spake Zarathustra" (1892)
Friedrich Nietzsche

"The purpose of computing is insight, not numbers"

-- "Introduction to Applied
Numerical Analysis" (1962)
Richard Hamming

Probably the German philosopher and prolific author's most famous quote, it was also uncomfortably prescient. For Nietzsche, like a disturbing number of historical genius-level intellectual figures, ended his career by going irretrievably stark, raving mad.

I guess he, himself, stared into the Darkness infinite a little too long, the poor devil.

\*   \*   \*   \*

"Don't get me wrong, Don Juan," I protested, "but I also want to know everything I can."

"No!" he said emphatically, "What is the sense of knowing things that are useless?"

-- "Don Juan: A Yaqui Way of Knowledge" (1968)
Carlos Castaneda

It was to the answer to a question I had not even thought of. Yet.

The mainstay of the U.S. ICBM land-based nuclear force for the Cold War after 1962 was the Minuteman, which evolved over 10 years from the Minuteman I in 1962, to the Minuteman III in 1970, the MM III being upgraded substantially in 1979.

The Minuteman was a marvel of technology: cheap, reliable, efficient, mass-produced mega-death. It was made possible by three things: the development of powerful solid fuels (replacing the problems of liquid fuels) for long range missiles, very light H-bombs, and higher precision Inertial Guidance Systems.

The first Minuteman I, ICBM missile went operational in 1962. By 1970, the Minuteman I had evolved to the increasingly more accurate and sophisticated Minuteman 2, and then the even more accurate and longer range Minuteman 3, which was upgraded in 1979 to the NS20 inertial guidance system, and double the yield with Oralloy secondary-equipped W78 warheads.

With only 25 minutes flight time for an incoming Soviet missile, U.S. ICBM's would be prime candidates for any technology that would automate and speed launching to carry out U.S. targeting plans as effectively as possible while under attack by a Soviet First Strike, before they were vaporized.

This was accomplished in two ways, beginning in 1971 with the DSP (Defense Support Program) constellation of 22,400 mile (35,800 km) high, geostationary orbit, infrared telescope satellites to reliably detect the launching (from the high temperature rocket plume) of the Soviet Rocket Forces' ICBM's, and giving a solid 25 minute warning of their "landing" in the U.S.

The second method was a revolutionary, new, and sophisticated Minuteman ICBM computer network to detect and reconfigure the retaliatory ICBM force when U.S. missile sites were knocked out by nuclear strikes.

This chapter is about the latter.

*　*　*　*

> "[A]fter all, the greatest and most calculating of
> killers is the national state, and this is true not
> only in international wars, but in domestic conflicts."
>
> -- "American Violence: A
>      Documentary History" (1970)
>      Richard Hofstadter

What was needed was a command and control communications network linking all the ICBM LCCs (launch control centers), and on to the individual missiles, stationed in their LF's (Launch Facilities).

It was to have three brand new network attributes:  reliable and speedy digital signaling, redundant (duplicated) links, and packet messaging, a new and efficient computer protocol.  More about this later.

It was a quote I had seen and read many times in Internet UseNet (founded in 1979) messages posted over the years.  It was a comment from well known, respected, and famed Internet icon John Gilmore (fifth employee of Sun Microsystems, creator of the alt.* UseNet hierarchy -- in the pre-web Usenet newsgroups -- one of the founders of the Electronic Frontier Foundation, etc., etc.), that he made in the early years (late 1980's), as the issue of censorship of UseNet groups or posts continually came up over and over again.  He had merely said:

> "The Internet interprets censorship as damage, and routes around it, "

His remark was quoted in a 1993 "TIME" magazine article, and a "sound bite" was born.  His statement was henceforth quoted endlessly, Gilmore's reputation – recognized as iron-clad and unassailable – adding to the technical validity of his comment.

I had heard the quote numerous times over the years, beginning in early 1992 when I first got on the 'Net, when the workstation network administrators at Bell-Northern Research secretly from upper management, hooked us engineers up to the marvelous new world.

> [BNR was the precursor to Nortel Networks, the telecom hardware and software
> manufacturer that went from multi-billion dollar Canadian multinational to bankrupt

in 2010 thanks to the foresight and leadership of incompetent CEO, John Roth, who around 1999 (7 years **after** his engineers were playing on the Internet) had "discovered" the 'Net, which he bragged to a reporter about.]

John Gilmore's clever line referring to the Net's architecture of computer hooked together to several other computers, that made for multiple, redundant links that provided the ability to resend message traffic along alternate paths, should a "node" – a computer -- in the fastest, direct network path "go down", "crash", break, become too congested/swamped with traffic, or otherwise become unavailable.

Sometimes it's not even necessary to hide the secrets. They're right in front of your face, and you don't notice them. Gilmore's remarks harkened to the origins of the Internet in the original computer network, the U.S. Department of Defense's ARPANet.

ARPA, the DoD's Advanced Research Projects Agency, had funded the first linking of different computers, in different cities, into a network, which eventually evolved into what became the Internet. Each computer was connected to a Digital PDP-8 computer (ancient!) which was itself connected to the transmission line network. The small computer, the PDP-8, was called an I.M.P. (Interface Message Processor), and performed the functions of dial-up, error checking, and message retransmission, routing and verification.

Due the redundant (multiple backups) links of the architecture, ARPANet was supposedly "nuclear survivable" – if during a nuclear attack a given computer was taken out by a nuclear missile, the other computers could still communicate, because the message traffic would automatically reroute itself to bypass the vaporized computer.

One day, this hit me like a flash, and suddenly everything was clear.

There was more.

There would be only one penultimate purpose for a computer network that could survive damage from a nuclear attack: a computer network sending launch orders or targeting updates to the 550 MM III Inter-Continental Ballistic Missile silos scattered across the Midwest, from Kansas to North Dakota. A computer network that kept track of each ICBM's targets and their priorities for nuclear destruction of Soviet targets.

This chapter is the story of how the Internet -- the Net -- evolved from a small Department of Defense test network that was to be the "proof of concept" for a national network linking the Minuteman III missile silos, and allow for the automatic dynamic re-targeting of the remaining -- unvaporized -- functioning missiles while under a Russian nuclear missile attack.

"Dynamic" re-targeting; it means "reconfigurable". It means "changeable in real-time in response to changing conditions". It means a current, up-to-date knowledge of remaining ICBM missiles, constantly updated in real-time by an "A-OK" signal from the computerized missiles, which were then kept constantly updated, if necessary, with a prioritized list of Russian targets, that would re-target the missiles in a nuclear war according to the balance of military priorities, with remaining, available missiles.

If a U.S. missile silo was hit by a Soviet nuclear missile warhead and destroyed, another missile silo could be reassigned the destroyed missile's targets. Basically the entire U.S. ICBM land-based force would be networked for dynamic targeting updates during a limited or all-out nuclear attack. [ADC 020813 (1979)] [DNA-6147T (1982)]

\* \* \* \*

"Death's at the bottom of everything, Martins.
Leave death to the professionals."

Searching my nuclear document collection for confirmation revealed that a "Command Data Buffer" (CDB) computer device was installed on the first MM III base in February 1975. Another document stated that the MM III force was upgraded with the CDB starting in 1973 and completed in 1977, and the older, single warhead Minuteman II force from 1977 to 1979.

Further, the CDB allowed for re-targeting of the missile in "25 minutes", revealing perhaps, the maximum time for manual re-programming the last of the hundreds of MM III ICBM's, or concealing a much lesser -- and classified -- time with installation of an Internet-like packet-switched -- automatic when under attack -- national computerized, all-out nuclear war, missile re-targeting system.

Command Data Buffer installation was completed on the first MM III base in February 1975. It allowed for re-targeting of MM III missile in one hour (as opposed to up to 36 hours in the older MM II). [Cochran (1984)] with the MM III on-board D37 computer. With an upgraded computer a single MM III could be re-targeted in about five minutes, a flight of ten missiles in an hour, and a full wing (150 missiles) in two hours.

> "Rapid re-targeting capability for Minuteman [MM3] to improve [under attack]
> strike planning capability." ["U.S. Strategic Objectives", p. 52 (1971)]

Additionally, each squadron of 50 LFs and 5 LCCs were interlinked, so that any one LCC could control any of the other 50 LFs ICBMs. The data rate even today is 1.3 Kbps, equivalent to three 56 kbytes per second (56 kbytes per second is the standard rate for a single telephone line). [Adkins (2006)]

\* \* \* \*

> "I don't think there's any point in being Irish
> if you don't know that the world is going to
> break your heart eventually."

> -- Daniel Patrick Moynihan (1963)
> on JFK's assassination

It was Plato, of the city-state of Athens in ancient Greece, and the most famous philosopher of all time, who realized it over twenty-three centuries ago, when he wrote: "Only the dead have known the end of war."

But was he signifying war's inevitability? Was he merely indicating its intrinsic part of humanity's behavior? Or was it just his observation of the facts of his time, history of time's past, with an extrapolation of a predicted future? Which I suppose are three different ways of saying the same thing.

Was it a statement of finality? Or a message of warning -- an invitation of the difficult journey we would have, to identify and begin the inner change necessary to end our instinctive need for the regular, organized, and systematic collective violence and mass slaughter, commonly known as war?

And all the side effects of our acceptance of warfare, with random regularity, at the drop of a dime, that must poison man's collective and individual behavior even when not engaged in warfare.

But I'll tell you one thing. It is so misguided as to call into question the basic logic, thought processes, and maturity of thinking of an ordinary person who believes revoking the private right of a citizen to own a rifle, or such, is anything useful, positive or meaningful in a world where everyone else, but the citizens of some democratic countries -- and all dictatorships -- are armed to the teeth with all manner of firepower and fearsome weaponry.

It's a simpleton's simple solution to a complex problem.  Or the solution of someone with the questionable motives of the hidden agenda.

\*   \*   \*   \*

"These in their dark nativity the Deep,
Shall yield us pregnant with infernal flame."

-- "Paradise Lost"
John Milton

I tried to study the ways and means -- the technology -- of the death-lovers, jingoistic, chauvinistic war-mongers, to understand war.  For a good three decades.

Perhaps I now understand it too well.

\*   \*   \*   \*

" 'God knows, I lie down to sleep so often with the
wish, sometimes with the hope, that I shall not
wake again.' "

-- "The Sorrows of Young Werther" (1773)
Johann von Goethe

In 1936, the brilliant British mathematician Alan Turing came up with the idea that was the theoretical basis that laid the ground-work for the programmable digital computer.  Exposed as a homosexual in 1952, in less open times, he was criminally prosecuted and cashiered in disgrace.

Turing committed suicide in 1954 from cyanide poisoning -- a poisoned apple.

(A reference to his favorite movie, the 1937 Walt Disney animated movie, "Sleeping
Beauty and the Seven Dwarfs" – based on the classic fairy tale  published in 1687,
made into the Tchaikovsky ballet first performed in 1890 – and allegedly the inspiration
of the PC computer giant, Apple Computing's, corporate name, first computer name,
and its corporate logo – an apple silhouette with one bite taken out of it.)

But Turing's idea for digital computers lived on, and changed the world.

So it goes...

The first digital computer was actually the Colossus, developed in the U.K. in 1943 for code-breaking purposes.  The first U.S. digital computer, named ENIAC, was built in the U.S. in 1946 for military purposes:  automated artillery table calculations, and came to be used for the design of atomic bombs.  The computer weighed 60,000 lbs. (27,000 kg.) and contained more than 18,000 vacuum tubes, and 1,500 relays, 10,000 capacitors, and 70,000 resistors.  [Fitzpatrick, p. 119 (1999)]

By the 1960's, with the introduction of individual transistors, computers were much smaller, but still massive in weight and volume, and were known as "mainframes", manufactured by industrial giants like I.B.M. for corporate and government use.

The idea of small personal computers had not even occurred to mainframe manufacturers, even if it was technically feasible or financially cost-effective, both of which it was not. Home PC's would have to wait for the 1970's and evolution of transistors to the integrated circuit (IC), which put many transistors and associated components on a single -- and tiny -- chip.

By the 1980's the price of the IC CPU had come down enough to begin the flourishing of the home PC. First came the IBM Personal Computer, or "PC", as it was quickly dubbed. The spread of home computing far and wide was due to cheaper clones that the PC wrought. The "accidental revolution" was about to begin. It was a technological revolution only, I'm sad to say, but so it goes.

*   *   *   *

> "[I]n Italy for 30 years, under the Borgia's, they had warfare, terror, murder, and bloodshed, but they produced Michelangelo, Leonardo da Vinci, and the Renaissance. In Switzerland they had brotherly love – they had 500 years of democracy and peace, and what did that produce? The cuckoo clock."
>
> -- Harry Lime
> "The Third Man" (1949)
> Graham Greene

But first there came a military need for small computers. Small -- in size, and particularly weight -- computers were needed for guidance control of the submarine-launched ballistic missiles (SLBM's) -- the Polaris Fleet Ballistic Missile -- under development starting in the mid-1950's for deployment in 1960.

Fast-forward to 1969.

Man's first landing -- and walking -- on the moon was by Neil Armstrong, with two other astronauts flying the Apollo 11 Saturn V three-stage, liquid-fueled booster (payload-carrying) rocket. The Saturn V was the largest and most powerful rocket ever constructed -- to generate the "escape velocity" necessary to break free from earth's powerful gravitational pull. The rocket needed to be strong enough not just to get into earth orbit, but for the extra "kick" necessary to break out of earth orbit, and into outer space -- heading towards the moon.

The date was 20 July 1969, and my father allowed me and my brother to stay up past midnight -- and way past our bedtime -- to watch on TV the fuzzy, murky image of man's first step on the moon by Armstrong.

But the technology base of the civilian Space Program was derived entirely from the then mostly secret -- or otherwise hidden from the public -- military **liquid-fueled** Inter-Continental Ballistic Missile (ICBM) program that started deployment in October 1959 in the U.S. -- with the Atlas missile, almost immediately trumped in 1962 by the introduction of the reliable, storable, quick-launching Minuteman I. And within eight years the solid-fueled missile had reached its first pinnacle with Minuteman III land-based, underground silo ensconced missile in 1970.

But for comparison purposes for the Guidance Control Computer, it is easy to take, for instance, the Apollo Guidance Computer (AGC). It was designed and developed at the MIT Instrumentation Laboratory, led by ICBM guidance expert Charles Draper, the details worked out by designer Eldon Hall. The contract for the development of the AGC was awarded to them in 1961 – a 5 year development effort, since the first AGC flew in an Apollo 3 test flight in 1966.

The AGC design drew heavily from the Polaris Submarine-Launched Ballistic Missile (SLBM) guidance system, also developed by (Eldon Hall of) the Draper group, and thus makes for a good comparison of the design and hardware of SLBM/ICBM guidance computer technology.

* * * *

"Having wandered some distance among gloomy rocks, I came to the entrance of a great cavern, in front of which I stood some time, astonished and unaware of such a thing. ... [A]nd after having remained there for some time, two contrary emotions arose in me, fear and desire -- fear of the threatening dark cavern, desire to to see whether there were any marvelous things within it..."

-- "The Notebooks of Leonardo da Vinci" (ca. 1475)
Leonardo da Vinci

The Apollo 11 moon-landing mission had one astronaut placed in orbit around the moon -- the Command Module --which released a smaller LEM -- Lunar Excursion Module -- to land on the moon with two astronauts, and when their moon walk and lunar sampling was complete, then blast-off to rendezvous -- rejoin -- the Command Module for return to Earth.  The AGC computers -- one on the CM and one on the LEM -- were used for "critical mission events" -- such as the calculations of thrust for a soft landing on the moon and the even more precise calculations for rendezvous of the returning LEM with the Command Module, the first time such a feat was ever attempted in space.

The AGC was needed for when the crew of the Apollo could not rely on earth ground station state computers due to the time delay for transmission from earth to moon and from moon to the earth -- a double delay totaling almost 3 seconds for *each* question, data (information) transfer, for calculation requests and answers from earth to reach Apollo.  Though seemingly a manageable delay, it was too long a delay for the necessary precision control for LEM space maneuvering where control measured in inches was necessary, as was required for LEM to CM rendezvous re-docking.

The Apollo system consisted of three major hardware pieces:  the inertial guidance unit, an optical space sextant system, and the general-purpose digital computer -- the AGC -- programmed to work with the data input by the other two hardware pieces, and command queries punched in by the crew.  Up till the development of these ICBM/SLBM computers, they had been massive pieces of hardware.  These missile computers were tiny in comparison, heralding in the early 1960's what in 1980 would be the PC for civilian mass market use.

The inertial guidance unit -- using three gyroscopes to stabilize a platform connected to  three accelerometer spacecraft acceleration measuring devices -- gave a continuous output of the Apollo's precise location in space, relative to earth, the spaceship's angle of movement (trajectory), and velocity.

The optical sextant made navigational angle measurements between a pair of stars, such as the sun and Polaris, the North star, and used to align periodically (error-correct) the inertial system platform based on star position, an innovation not used on the Polaris SLBM (though considered later for the Trident SLBM).

But by today's standards of the home or business PC -- the year 2008 -- the AGC of 1969 was a fossil. But its small size made it the antecedent of the PC.

The AGC weighed a hefty 70 lbs.(32 kg.), and had a volume of 1 cubic feet.  The AGC was a 16-bit computer and was based on one of the first military applications of the integrated circuit.  Today's computers have 64 bit CPU's (the Intel Pentium 4).

The clock speed of the AGC's CPU was 2 MHz . Today, the Pentium 4 has a clock/instruction speed of up to 3.8 GHz (a calculation speed of almost 2 million times as fast as the AGC) -- at only 115 watts of power consumption, compared to 70 watts for the much lower computing power of the AGC.

There was only 2 Kb of erasable (read or write) RAM memory, running on a cycle time of 12 μs. Today 2 Gigabytes of RAM is typical (500,000 times the amount of AGC RAM, as well as being smaller, lighter, and faster than the older technology Apollo's 4 Kb).

Hard drives didn't exist at the time, and ROM (Read-Only Memory) was faster anyway. The computer's program was stored in about 74 Kb of ROM (Read-Only Memory), the memory of the old magnetic core rope type. The AGC program allowed for only **34** calculations/instructions. Today, hard drives are used for permanent memory storage, and 500 Gb hard drives are common (almost 7 million times the amount of AGC ROM). Programs and applications (instructions) are many.

Input today is by alpha-numeric (alphabet-number) keyboard or mouse click, and output to the user is through a video screen. In the AGC, input was by the DSKY (Display and Keyboard), with the astronauts punching in one of the only 34 calculations/instruction, coded as 4 digits *numbers*. Output was also a number: three 5 digit number plus sign ("+" or "-").

There were 200 fixed input and output ports on the AGC: to the IMU, the crew display and control, the radar and optical systems, the various rocket and jet thrusters.

Crew commands input were, for example, "display gimbal angles" (of the inertial guidance system) or "compute abort velocity". The input (crew request) and output computer answer) was on a display and keyboard unit. Input and output was through coded numbers **only** -- a primitive user  non-friendly interface that saved weight, complexity, and was consistent with the level of computer advancement of the time.

*    *    *    *

"Have you ever seen any of your victims?"

"...Victims?  Don't be melodramatic.  Look down there. [at the
crowds down below, tiny moving dots] Tell me, would you really
feel any pity if one of those dots stopped moving forever?  If I
offered you twenty thousand pounds for every dot that stopped,
would you really, old man, tell me to keep my money, or would
you calculate how many dots you could afford to spare?

-- Martins to Harry Lime
"The Third Man" (1949)
Graham Greene

Pure research is about the mind, and its ultimate capabilities. It is on the edge of genius, or genius itself. It is about the mind being stretched to the extent of its true capabilities into the beauty of purely ideas and concepts. Gymnastics of the mental realm. Dreams of the mind awake.

Pure research is the stratosphere of scientific and technological development. Pure research generates new concepts as:

1) extensions or combinations of existing ideas or technology,

2) extensions of existing technology that require further technical -- hardware – improvement, advancements, or development before they can be implemented,

or

3) entirely new revolutionary ideas of developments unrelated to existing technology which may or may not require technological developments to catch up for implementation. Or that require human minds of those in positions of power to realize the importance of to accept, or find the need or the funding for.

After the lesson of WW2 of the importance of technology (radar for instance) to the successful -- and easy -- prosecution to victory of warfare, a structure and administering bureaucracy was erected in the U.S. Pure military research became the realm of an assortment of well-funded groups, such as -- most famously -- the RAND Corporation, as well as Princeton's Institute for Advanced Studies, the Stanford Research Institute (SRI), the Institute for Defense Analysis, the JASON group of experts -- associated now with the MITRE Corporation -- and finally the government agency overseeing and financing such R&D, ARPA.

Before ARPANET, which arose between 1969 to its public unveiling in 1972, Paul Baran in 1960 began working on a DoD network piggy-backing on the many AM stations, then in existence. The military worked in HF (High Frequency) frequencies, that the temporary destruction of the ionosphere by a nuclear bomb would put out of action. So, the low frequency ground wave of commercial AM radio stations were it. And the radio stations could relay its messages from station to station. Paul Baran's idea was rejected, but it was given to the Rome Air Development Center for further development.

Rome got it up and running as a teletype system, running as a network between 12 AM stations. The piggy-back was at the low frequency of the AM station's 20 Hz signal, the DoD superimposed an FM signal, the whole thing working perfectly, and undetectable by the AM listener. In 1960 at RAND, his work continued. With a number of collaborators, they built an experimental net that showed that if you had a redundancy of three, you had a very robust network. So each computer site would have to ideally connect to at least three other separate nodes (computers). [Brand (2001)]

ARPANET, the first wide-area computer network -- connecting distantly sited computers together so researchers could communicate, use each others programs remotely, and exchange information -- was conceived, funded, and launched in 1969, but had fewer than a couple of dozen computers -- "nodes" -- connected together and communicating by early 1972 -- the first long distance computer network. Indeed, it was the first long distance functioning computer network *ever*.

The ARPANET -- built under a concept put forward by ARPA -- connected together a bunch of the large university mainframes in existence at the time, linking them together in a national network. And the simplest, quickest, cheapest, and easiest way to get this done was to piggy-back on an already existing national network -- the long distance telephone network.

So, they just rented a bunch of Bell Network long distance toll lines -- "trunk lines" – connected them to new specially designed routing and message translation computers (called "IMPs" -- Interface Message Processor – and using a common language TCP, Transmission Control Protocol), and connected them to the different large mainframe computers – which "spoke" different computer languages -- at the various universities and institutions recruited for the "experiment" -- an ever-growing network, as more and more computer mainframes signed up after the concept was debugged of most of its initial -- and inevitable -- hardware and software problems, and stable and up and running.

Each computer would communicate with the IMP, which would translate each individual computer's language into the language of the network; each computer connected to one or more other computer's and would route messages for different computer's so that each computer would not have to be connected to every other computer. Used long distance telephone lines; re-send messages if one was lost or garbled on transmission on the low quality lines.

Bell uses a protocol called circuit-switching: you need to use a dedicated connection for the whole voice conversation. Computers don't need such a dedicated link. Computer communications are usually in occasional bursts of data -- they are "bursty".

Thus "packet switching" was invented (thought of) in the early 1960's, first by Paul Baran at Rand Corporation, in the U.S., and independently, and almost simultaneously, by Donald Davies at the National Physical Lab in the United Kingdom, and who coined the term. Packet switching greatly increased the efficiency of computer-to-computer communications over the very inefficient circuit-switching.

Packet-switched. Break up the message up into individual 128 byte long segments (one byte = 8 bits = 1 alphanumeric character), then add a header (for the destination) and the sequence (the order) number of the packet, and finally, a check digit for the receiving computer to detect a corrupted packet (which results in the destination computer sending a "re-send packet message" back to the originating computer.

And typically with there being multiple links between different computers in a large, wide-area network, packets can take any of multiple routes via intermediate "nodes" (other computers on the network) to the final, desired destination computer, avoiding "down" (out-of-service) nodes or busy nodes, and taking advantage of spare link capacity (spare "bandwidth") to speed message transmission.

Each computer on the computer network might be different – a different manufacturer, or model. Enter the IMP (Intermediate Message Processor). The IMP connects each computer to the network, and translates the messages it is sending out to a common "language", and translates received messages into the computer's own language that it understands.

The first IMP, based on a modified Honeywell 516 mini-computer [Reed, p. 20-7 (1990)], was delivered to UCLA in mid-1969, and it was used to hook UCLA to three other sites, in what became a four "node" network: UCLA, the Stanford Research Institute (SRI), UC Santa Barbara, and the University of Utah, Salt Lake City.

By October 1969, the first connection had been setup between SRI and UCLA (which are in the same city). An operator began to type "login" to setup a connection with UCLA, and at "log", the IMP crashed. The first "bug" in ARPANET had been found!

It was fixed, and the ARPANET became operational. By 1971, there were twenty-one nodes in the initial ARPANET, with thirty different university sites to be funded from across the United States, including MIT, Harvard, Carnegie Mellon University, and the University of Illinois. [AD-A528 970 (2005)]

"Distributed computing" was demonstrated, and the Internet was starting to take off, in spite of the cost of IMP's in the early 1980's being around $50K in 1980 dollars. The original ARPANET was split off in mid-1975, and transferred to the Defense Communications Agency. Working with the NSA, they developed a "private line interface" (PLI) to encrypt classified messages. One PLI would be on each end of the data "conversation". (The successors to the PLI were the IPLI and BLACKER.)

\*   \*   \*   \*

"Countries don't mistrust each other because of
armaments they build up, they build up armaments
because they mistrust each other."

-- Salvador de Madariaga (ca. 1936)

The network was the first implementation of a new type of communication type -- "protocol" -- called "packet switching" which was ideal for a computer network, as opposed to the only other type of network, telephone networks which used "channel switching". Channel switching requires the network to set aside a dedicated channel for the sole use of the two telephone users until they are finished and hang up. Though necessary, channel switching is very inefficient. It is estimated that up to 50% of a typical voice conversation is actually silence.

The idea, on the other hand, of packet switching is this. Computers are very fast, and humans are by comparison, very slow. The computer data transmission is thus "bursty" -- it is transmitted in an occasional burst by the computer as the human operator presses the "Return" key. Thus it doesn't need a dedicated channel – which would be unused most of the time by the single computer. Thus it was realized than many computers can use the same transmission line.

Thus came the idea – developed conceptually, independently and almost simultaneously, in 1962 by Paul Baran in the U.S., and Donald Davies in England -- of a computer network based on "packets" of computer data sent down the pipe -- transmission line -- as opposed to having a dedicated -- and wasteful -- "channel", as in the voice communications telephone network.

Further, the concept worked thusly. A computer message is broken down into many individual, fixed size, numbered smaller messages named "packets". They are then transmitted to the network, where individual packets take any route on the network, and are passed from network node to network node, until they reach their destination, where they are reassembled and ordered correctly into the original full message.

The packet network is so efficient because, as stated previously, unlike a telephone network, computers don't require a dedicated path -- the setting aside of a continuous, and expensive separate channel for communication, since they only communicate in the occasional short, quick burst of messages.

\*   \*   \*   \*

> "Security can now only be achieved in common.
> No longer against each other, but only with each
> other [together], shall we be secure."
>
> -- Egon Bar (1981)
> West German Bundestag member
> speech

In essence, the Internet was inevitable once computers became cheap and widespread. The concept of two users sitting at separate computers -- connected by an electrical cable across the room -- talking to each other is trivial. The work to make it happens is simply the details, the purpose, and the sophistication of what you want to do.

If you were an engineer working for a big corporation, professional chemist, physicist, or undergraduate science student who hung around the computer room around the late 1980's or early 1990's, and paid attention to the goings on, you were likely witness to a technological sunrise that was amazing to watch.

I was one of them -- a female colleague turned me on to "the InterNet" in 1992, when I sat beside her in a long-forgotten company course we both took and showed me this amazing new thing on our Unix operating system computer "workstations" that changed the course of my life, by showing me a strange new world, that was being born, and made my life suddenly quite a bit more interesting.

That was one of the really cool days of my life, and I don't even remember her name.

What she had showed me on a UNIX workstation -- a $30,000 computer that I did my work on at Nortel, in Ottawa, working on telephone company telecommunications software for the gigantic computers that ran the public telephone network. What the fellow engineer had showed me was a Unix application called "Usenet", that allowed someone to "post", read and respond to messages on any of hundreds of topics ranging from computer software topics, hardware problems, to recreational groups -- guns, archery, weight-lifting, amateur photography -- to the frivolous -- jokes, cult movies, politics, both serious and outrageous.

Most are long gone now -- having moved on to bigger and better things --, but when I was new on the Net, you could still see messages from the famous, and important -- the people who managed the net, were the gods of this expanding new universe, or revered names of "net gods" and "net personalities" who no longer posted, but who were still remembered by thousands, and honored silently by those thousands who knew their reputations, genius, and contribution to this, that, and the other thing -- all computer network-related.

And there were colorful characters, too. "Net kooks", eccentrics, the gregarious, and the entertaining.

And there were on-going controversies. Whether gay discussion groups should be allowed. "Yes" demanded the libertarian-types, as well as the "Live and let live", types. (And thus the response of John Gilmore quoted at the top of this chapter about "routing around censorship".)

"No", said the bigots, the rigid, the anti-gay, and the Christian fundamentalists, "Who cares?" said most, along with the ever-popular, "If you don't want to read a gay message group, don't read it. But what give you the right to prevent someone who wants to read a gay message group if they *want* to?".

A system [local network] administrator -- sysadmin -- **that's** who.

A sysadmin using his power to unethically impose his prejudices on the users on this computer network, or to cowardly avoid controversy. But this is the real world, and it's a gutsy user who's going to complain to his boss that he wants to read a gay newsgroup, but the sysadmin is censoring it...

The controversy raged endlessly. Ain't freedom of speech -- the First Amendment of the U.S. Bill of Rights -- wonderful?

*　*　*　*

> "I wonder why I live alone here,
> I wonder why we spend these nights together,
> Is this the room I'll live my life forever,
> I wonder why in L.A., To live and die in L.A.
>
> I wonder why we waste our lives here,
> When we could run away to paradise.
> But I am held in some invisible vice,
> And I can't get away, To live and die in L.A."
>
> -- "To Live and Die in L.A." (1985)
> Wang Chung, Jack Hues

Arpanet began "a DoD-wide implementation...from 1975 onwards." "Notable early contributions had been made by P. Baran and collaborators at RAND. Baran's work in the early 1960's outline  a distributed, survivable digital system for the Air Force..." [Reed (1990)]

*　*　*　*

[Distilled from "ICBM Security Classification Guide" (1997), regarding Minuteman III and MX/Peacekeeper ICBM's]:

Command Data Buffer

LCF Processor (Launch Control Center)

Encryption and Authentication: cryptographic input keyer KSK-45, to transfer the keying variable into the COMSEC TSEC (Transmission Security codename) KIK-45. One KIK-45 holds half of the KI-45 key, combined with USKAU-100 A/B keying variable assembly. COM SEC-classified code-book stored at LCC.

A Launch code required to launch missiles, which are radio-controlled for launch.

For the Airborne Launch Control System there is an RF Amplifier, tuned to a classified frequency, probably in the HF (High Frequency) band.

The RMP (Rapid Memory Processor) (using RAM memory) processes Emergency Action Messages (RAMs). EAMs direct the war and may generally contain or be used to determine necessary data, such as authentication codes, time values, and targeting support information. They are bit level messages with characteristic wave-forms, and character coding schemes such as PLSO and STUTTER, with error-correction encoding.

VDU (Video Display Unit)
WSP (Weapon System Processor) with RAM (volatile) memory
VAX-11/750 Computer, using the VMS operating system

Strategic Target Planning:

ICBM target cases (SIOP-ESI)
      Operational targeting program tapes        )
      Master Launch Facility (LF) data base tape,    )
          encrypted                      )-- two person rule in effect
      MOTP3 data base tape, is targeting database  )
      Execution Plan Program (EPP) database and   )
          message analysis (SIOP-ESI)        )
      Strategic Targeting Support Software (STSS) Programs

The Trajectory and Missile Parameter (TAMP) provides these parameters to the MM3.
Console Operations Program (COP) and Database, with firmware, and Boot ROM, is the main    Weapon System Processor (WSP) program, providing command, control, and communications.
TCI/EPCI Database (Target Case Info/Execution Plan Case Info) (SIOP-ESI).
Force Direction Message (FDM) Format DB is the database for force direction messages.
Operational Ground Program Offload Tape, transferred, and run by the MM3 fight computer.

The Launch Facility (LF) is interrogated by the Missile Alert Facility (MAF).

The Message Processor (MP) performs rapid message processing functions for SIOP messages, coming from different backup HA (High Altitude – satellite, and airborne command posts) communications systems. The Transmission Integrator (TI) prepares acknowledgment messages for transmission to different HA communication systems.

Geodetic (straight line trajectory to the target), gravimetric, astronomic, climate, atmospheric, and geometric parameters are some of the data required for missile launch. North American Datum Coordinates: degree, minute, and second. Absolute acceleration due to gravity (nominally 9.8 m/sec2 at sea level), and height above mean sea level.

The constantly changing atmosphere causes measurable and potentially significant deviation in the trajectory of RV's. Atmospheric perturbations during reentry may be estimated by the equation:

Rate of deceleration = rate of change of velocity = $dV/dt = -(1/2) g \rho Va/(W/C_D A)$, where:

                $g$ = acceleration of gravity (9.8 m/s2)
                $\rho$ = density of air

V = speed of vehicle
A = reentry path angle
$W/C_D A$  ( = β, the ballistic coefficient, the weight-to-drag ratio)

[AD353 247, p. 7 (1964)]

"Then we're stupid; an' we'll die."

> -- Pris
>  "Blade Runner" (1982)
>   Hampton Francher, David Peoples

# "A Screaming Comes Across the Sky"

## Point-to-Point:  The Inter-Continental Ballistic Missile

"Things will be better when everybody's gone.
... When we're all gone at last then there'll be
nobody here but [D]eath, and his days will be
numbered too.  He'll be out on the road there
with nothing to do and nobody to do it to. ...
And that's how it will be."

>                    -- old man
>                     "The Road" (2006)
>                     Cormac McCarthy


"**We** don't belong here? ... On the contrary, I belong
here completely and utterly.  I'm home.  You, with
your absurd notions of a perfect and harmonious
society.  Drivel.  The world has caught up and
surpassed me.  Ninety years ago, I was a freak.
Today, I'm an amateur. ... The future isn't what you
thought.  It's what I am. ...

[H.G. Wells roughly grabs Jack the Ripper, and then
realizing his mistaken outburst, releases him.]  "Stop it!!"
yells Wells.

"It's catching, isn't it?  Violence."

>                    -- Jack the Ripper to H.G. Wells,
>                     having time-traveled to 1979
>                     "Time After Time" (1979)


The end will come with a flash streaking across the sky.

At night or even in the brightness of a sunny day, the super-heated carbon skin of the MIRV reentry vehicles glowing with incandescent white light, will be visible as they descend on the final leg of their voyage from Russia, over the North Pole and dropping down to their targets across the U.S. mainland.

As the ICBM missiles burn out and fall away, the last phase of unpowered trajectory takes them from outer space, and as the RV's descend to an altitude of 46 km (29 miles) their carbon skin will begin glowing from atmospheric friction with the air.  First bright red, then orange, then blue, finally maintaining a glowing, bright, incandescent, white light.

At a height of between 46 to a minimum of 23 km (29 - 14 miles) in height, and below as they descend, the RV's will be clearly visible as a white flash of light falling from the sky for about 6 seconds, brighter than sunlight [AD-A229 778, p. 5-25 to 5-34 (1990)], followed by the overwhelming blinding flash of the nuclear explosion as the warhead detonates over its city, or on top of its ICBM silo target.

World War 3 will have been brought to America, with over a thousand such explosions, as its cities are incinerated along with hundreds of millions of deaths. The USSR and China will experience similar fates when America retaliates.

It will be called "the Day" by the survivors.

*   *   *   *

"The plunge of civilization into this abyss of blood and darkness...is a
thing so gives away the whole long age during which we have supposed
the world to be, with whatever abatement, gradually bettering, that to have
to take it all now for what the treacherous years were all the while really
making for and *meaning* is too tragic for any words."

-- "The Letters of Henry James" (1920)
Henry James

"A screaming comes across the sky. It has happened before, but there is nothing to compare it to now."

It's the first line of Thomas Pynchon's "Gravity's Rainbow", the second best fiction book in the history of English literature. The devastating V-2. It is about the future when the Nazis have conquered all of continental Europe, and continue bombarding England with V-2's, with an almost 1 ton HE warhead. The story is about a low-ranking U.S. officer, Lieutenant Tyrone Slothrop stationed in England, whose locations of his many one-night stands is where, soon after, launched V-2's land, and explode.

The missile has a curved trajectory in the sky – like a rainbow – pulled down by gravity. And on heating by atmospheric re-entry, glowing like a prism, with all the colors from red to blue one after another, until finally radiating a brilliant white light. Ranging from the invisible Infra-red to Ultraviolet, and with no pot of gold found by leprechauns at the end of its descent from the sky.

Maybe evil leprechauns, exiled from Ireland...

The V-2 was the first supersonic, guided Short Range Ballistic Missile, so its velocity was faster than the speed of sound, and thus its arrival was silent, until it cratered its landing site, or brought down a cluster of urban homes or buildings, or kills civilians – whatever it hits. The newly invented coastal radar couldn't even see its arrival because they were aimed at German bomber level, and the radar couldn't reach that high, anyway. The maximum height (apogee) of the trajectory of the V-2 was something like 45 [AD-A272 447 (1993)] or 63 miles [AD-A434 478 (1961)].

The Sonic Boom was delayed by a second or two. So there was no audible advanced warning of the arrival of the supersonic air vehicle. The V-2, Vengeance Weapon 2, _Vergeltungswaffen-2_. It was truly a terror weapon. Instant, unexpected, mass death in the enemy's homeland: a "strategic" strike.

"The primary duty of the ICBM crew members is to be prepared to launch their missiles towards enemy targets when ordered by the President of the United States". A "task not to be taken lightly," says an internal 1988 handbook for ICBM launch crews. [Hodge, p. 107 (2008)]

It is generally agreed that the discovery of gunpowder (aka black powder) was in 9th Century China, by Taoist philosophers. Little did they know...

Rockets – unguided missiles – soon followed, first made by packing hollow bamboo stems with gunpowder. The first use of rockets in warfare was their successful use by the Mongol leader Genghis Khan against Europeans in the 13th Century.

The first **metal-cased** rockets – a significant advance – were used in India by the Haider Ali, ruler of the Kingdom of Mysore, against the invading British East India Company in 1792. The iron casing was eight inches (20 cm) long, and the rockets had a range of over half a mile (~1 km). Barrages of hundreds of rockets were fired against the British.

The British sent "dud"/captured samples back to the Royal Arsenal at Woolwich, London, where Sir William Congreve reverse-engineered the design, developing in 1805 the much more famous "Congreve rocket". It was used successfully in the Napoleonic wars.

The Indian and Congreve rockets were stabilized in flight by a stick protruding from the base of the rocket, familiar to those who have fired fireworks "bottle rockets". In 1844, William Hale, a self-taught British inventor, improved the design by slightly off-setting the rocket's thrust to impart a gyroscopic spin that stabilized the rocket, eliminated the need for the protruding stick, and allowing a doubling of the size of the 32 lb. (15 kg) Congreve rockets to the Hale design weighing 60 lbs. (27 kg). [Ford, p. 111-117 (2011)]

The theory of ballistic missiles began with the ideas of the "father of space travel", the Russian genius, Konstantin Tsiolkovsky. In 1883 he derived many of the mathematical formulae pertaining to rockets. He realized that rockets could travel through the vacuum of space by ejecting matter at high speed, and not from the ejected gases pushing against the atmosphere. It was simply an application of Newton's Third Law of Motion: for every action, there is an equal but opposite reaction.

Continuing his earlier work, Tsiolkovsky realized that a multistage rocket would be much more efficient than a single stage. His work was finally partially published in 1911, in the "Aviation Reporter". The multistage rocket was also patented in 1911 by a Belgian, Dr. Andre Bing (German Patent #236,377).

In 1935, at the age of 78, Tsiolkovsky died, having written more than 500 scientific papers and books.

In 1889, the Swedish engineer Carl de Laval made a device for gas turbines known as a convergent-divergent nozzle, that converted thermal energy more efficiently into kinetic energy – motion. It was perfect for rockets too. The gases produced in the combustion chamber had high pressure, but low velocity. However, escaping through a rocket nozzle, the exhaust gases increased in speed as they dropped down in pressure, providing increased thrust to the rocket in doing so. However, one of the most serious problem areas in solid propellant rocket motors is finding suitable materials for the nozzle throat, which are subject to high heat and erosion. [AD268 311 (1961)]

American Robert Goddard was granted U.S. Patents #1,102,653 and #1,103,503 in 1914. They covered multistage rockets, liquid propellant feed systems, and combustion chambers and nozzles. In 1919, now a professor at Clark University, the Smithsonian Institute agreed to print his 69-page treatise on rocket technology, including sections on rocket efficiency, and based on theory as well as experimentation.

In 1923, Hermann Oberth published a 92-page monograph, "The Rocket into Interplanetary Space", independently reproducing much of Tsiolkovsy's work.

Then there was synergy. Soon all three were in communication: Tsiolkovsy, Goddard, and Oberth, all mathematics teachers, had independently come up with the basic laws of rocketry in three different countries, at about the same time.

In 1925, Walter Hohmann published "Die Erreichbarkeit der Himmelskorper" ("The Reachability of Celestial Bodies"), a volume so technically detailed, it was still being consulted thirty-five years later by NASA scientists.

Dr. Robert Goddard, an American mathematics professor at Clark University in Worcester, Massachusetts had worked on his rocket for two years, before on December 6, 1925 he static-tested his small rocket. The propellants were LOX and gasoline, pressure-fed into the combustion chamber. On March 16, 1926 he launched the world's first liquid-fueled rocket. On March 28, 1935, he reached 2.25 km (7,500'), equaling the height achieved by the later German A-2.

The fundamentals of the rocket: the specific impulse of the fuel combination, fuel consumption, mass ratio, effective exhaust velocity, chamber pressures and temperatures. The thrust-time curve of the rocket engine.

First there was the development of a sophisticated, modern weapon: the liquid-fueled (alcohol and liquid oxygen (LOX)) V-2.

In 1932, an aeronautical engineer, Werner von Braun became the first employee of the German Army's missile development program. There was Captain Walter Dornberger, with a master's degree in mechanical engineering, and with a great deal of knowledge in artillery ballistics. Then there was Heinrich Grunow, a superb mechanic. Then Walter Riedel, experienced in handling LOX. Dr. Paul Heyland's company was the only industrial company in Germany that had the capability of manufacturing the desired missile engines. In January 1933, Dr. Kurt Wahmke, joined the missile team. He had written a paper on exhaust gases through cylindrical nozzles.

First they developed the _Aggregat_ 1 (A-1), meaning "Prototype 1". But they had problems with engine ignition and burn through, the engine combustion chamber and exhaust nozzle first being made of aluminum.

In the nose was a 30 kg (65 lbs.) flywheel, with its axis running on ball bearings: a gyroscopic flight trajectory stabilizer. The rocket engine was placed at the bottom, fed by the fuel tank – cylindrical with hemispherical ends – filled with 75% ethyl alcohol, and a thin-walled, open fiberglass container holding the LOX, placed within the top of the fuel tank. A small pressurized nitrogen gas flask forced the 38.6 kg (85 lbs.) of propellant into a new double-walled, steel regenerative cooling combustion chamber.

This was eventually changed to a lighter fuel pump method using a steam-powered turbo-pump, the steam generated by hydrogen peroxide mixed with potassium permanganate. It could handle the specification requirement of 129 kg (284 lbs.) per second of methanol and LOX. The hot steam had a pressure of 21 atmospheres (326 psi), turning a single-wheeled turbine only 47 cm (18.5") in diameter, at 50,000 rpm, with a power of 675 hp, and drove the turbine whell's impellers, mixing the correct proportions of LOX and fuel, and forcing the mixture into the combustion chamber.

But LOX was cryogenic, and though a much better oxidizer, it was expensive and hard to produce, store, and handle. It was slow to load, and had to be loaded immediately before missile launch. In March 1934, they changed to 90% hydrogen peroxide ($H_2O_2$) and alcohol in a steel tank, that was also fed into the combustion chamber premixed.

For missile flight stability, a gyroscope working on three axes would be required. And the shape of the rocket would be patterned on the German Mauser 8 mm infantry rifle bullet, which was aerodynamic, with a very low drag. The engine chamber was changed to the much more heat-conductive, heat-dissipating copper.

The A3 had 1,500 kg (1.5 tons) of thrust with a burning time of 45 seconds. Metal fins at the bottom of the missile casing, operated by a servo, steered the missile. But they only worked at high speed, and were useless at launch and the missile's initial low speed, where the air velocity they needed to work was negligible. The solution was to place a set of four control vanes inside the flow of the exhaust gas jet. The exhaust velocity of the combustion gases was almost 1.95 km/s (6,500 ft/s), remaining virtually the same during the boost phase, and thus a constant force from the very beginning. The V-2's velocity was twice that of the U.S. M1 Garand's heavy .30-06 caliber rifle bullet.

Originally molybdenum vanes were used, but these were replaced by graphite vanes at 1% of the cost, and better performance, since graphite has an even higher melting point than molybdenum. Copper-infiltrated tungsten was also later considered for use. [U.S.P. #5,082,202, Navy, filed 1975, declassified 1992].

Their place of research was Peenemunde, Germany.  The name Peenemunde means "the mouth of the Peene", since it's on the bank of the mouth of the Peene River, emptying into the Baltic Sea.

The German "Paris Gun" of World War I had a range of 125 km (78 miles) with a steel artillery shell containing 10 kg HE (22 lbs.).  The A-4 would have 100x the amount:  1 ton.  The spec was to build a rocket with twice the range of the Paris Gun, and a dispersion of 2 - 3 miles, far better than artillery, where a 50% dispersion at a range of 4 - 5 miles was acceptable.  In other words 50% of impact point should be within 2 or 3 meters per km of range, or 0.5 - 0.75 km within range.

The fuel pump was driven by a steam turbine, and was a dual pump cast in light alloy.  The steam was generated by using eight 7-litre compressed air bottles to pump and decompose into steam 80% hydrogen peroxide ($H_2O_2$) with potassium permanganate ($KMnO_4$); the two-stage 500 hp turbine drove two centrifugal pumps at approximately 4,300 rpm, pumping the alcohol and liquid oxygen at pressures around 350 psi.  The pumps were only about a foot in diameter.

The rocket motor was steel, six feet long, and double walled for cooling by the alcohol fuel.  It had 18 burners arranged around its top end, each with three rows of jets which mixed the alcohol and oxygen before burning them at a temperature of 2,700 °C.  [Johnson (1978)]

The supersonic wind tunnel they built a Peenemunde, was the best in the world.  At Mach 4.5 the shape of the fin problem was solved.

By placing three injection heads in the outer chamber it produced 4.5 tons of thrust.  Then 18 injection heads in two concentric circles at the base of the inner combustion chamber brought the thrust up to 25 tons.


The A-4 was a single-stage rocket.

A4 dimensions:  1.65 m (65") dia. x 14 m (46') high; Stabilizing fins of dia. 3.44 m (11.3') at the bottom of missile.
Fuel + oxidizer tanks:  62.15 m high; engine less than 4 m high; guidance:  7.5 m high.
Propellant:  5,533 kg LOX (tank volume 4.8 cubic meters) + 4,173 kg methyl alcohol (containing 25% water (tank volume 5.18 cubic meters) (made of aluminum and sat above LOX tank).
Range:  190 miles, which took a few seconds over 5 minutes flight time.
Warhead:  Conical forward housing; 2.1 m high, 850 kg amatol (60% ammonium nitrate/40% TNT) high explosive.  A central small steel tube filled with TNT pellets ran through the warhead to act as a booster and transmit the detonation of the fuse to the main charge.
Total unfueled weight:  4,000 kg (4 Metric Tons)
Total fueled weight:  13.5 Metric Tons
Pressure at exhaust:  15.2 atmospheres, down to 0.  Temperature:  2,360 °C. down to 1,315 °C.
Thrust:  25 tons thrust from burning up to 125 liters (33 gal.) of fuel and LOX per second.  Burning time was 65 seconds.
Velocity:  2,050 m/s.
Guidance:  Two gyroscopes (one for roll and yaw, and the other for pitch) controlling four internal carbon jet vanes in the exhaust stream, and four external vanes in the same plane as the four large rocket fins.  Also used a gyro-integrating accelerometer.  [Hogg (1999)]


Originally, it had a 0.091" thick 3S aluminum (Al-Mg) alloy as the forward warhead casing (with a 0.109" thick stainless steel for the rear warhead casing).  But this was replaced due to wartime shortages to a 6 mm (0.25") steel casing, with a slightly decreased warhead weight.

*   *   *   *

"As for [medium range strategic] rocket secrets, there

are none, either in theory or design.  All one needs is
a reading knowledge of English, French, or German,
and a good collection of recent technical journals,
patent literature, etc."

-- Prof. Martin Summerfield (1964)
Princeton University
NYT Letter to the Editor (8 Nov.)

The start of the evolution of ballistic missiles – to the ultimate, the ICBM, the solid-fueled Minuteman I, introduced in 1962 – was in 1944 with the WWII development of the first ballistic missile, the V-2, by the Nazis (and called the A-4 – prototype 4 – by its military designers).

It was a supersonic missile – liquid-fueled and inertial-guided – flying at almost four times the speed of sound, which meant the missile's arrival preceded its sound, making it all the more terrifying when it struck its target silently with no warning sound, out of the blue.  But as a novel, completely new technology, it was a short-range ballistic missile (SRBM) with a range of only 200 miles.

Over 1,500 were launched at London, and 1,100 reached their target, causing 3,000 deaths, and 7,000 casualties (a fraction of Allied bombing raids casualties on Germany).  And that was the reason it was named the V-2, for  "Vengeance 2" (the follow-up to the earlier, simpler, much slower, and less effective V-1, the first "cruise missile").

Its use as a tactical military weapon occurred when in mid-1944 1,600 missiles targeted the port of Antwerp, Belgium, through which Allied forces (who had already successfully established a bridgehead in Normandy on D-Day of June 6th, 1944) were bringing in additional soldiers and equipment.

Intelligence reports from agents, confirmed by the famous deciphering of the German ENIGMA code, along with airplane surveillance, and then finally photo-reconnaissance in June 1943, had confirmed the testing and manufacturing site to be Peenemunde, a remote island off the German coast in the Baltic Sea.  In August it was hit by a squadron of 597 Lancaster bombers (41 of which were shot down), severely delaying, but not stopping production, which was nonetheless moved to Blizna, in occupied Poland, to outwit Allied intelligence.

In late 1936, Dr. Walter Thiel, the leading expert on rocket engines, had joined the A-4 design team.  He developed a nozzle through which the premixed fuel and oxidizer were injected as a fine vapor into the combustion chamber, instead of separately, producing higher performance, and shortening the engine combustion chamber down to 30 cm (1') long, which increased the exhaust velocity to 2,070 m/s (6,900 ft/s), close to the maximum theoretical value of 2,250 m/s (7,500 ft/s).  The fuel consumption was also reduced to 4.5 g/s per kg of thrust.  But they had to scale up for the A4.  Burn-through of the nozzle was solved by cooling with a thin film of cold alcohol fuel, just before it entered the nozzle.

Prof. Hermann Oberth and others had calculated that the maximum thrust was obtained when the optimum expansion of exhaust gases was reached when the nozzle exit pressure equaled the ambient pressure

So far all fin-stabilized rockets lost their stability at supersonic speeds, and supersonic wind tunnels weren't in existence.  To test the rocket's inherent stability, therefore, without any stabilization, the aerodynamic stability was measured by dropping models from 6 km (4 miles), reaching a maximum supersonic speed at 900 m (3,000') of 1,300 km/hr (800 mph).  Rotary oscillations never exceeded 5°.

Boykow's' acceleration-measuring, double integration device corrected any lateral deviation caused by the wind.

The recently developed 85% $H_2O_2$, decomposed to water, oxygen and heat, to create steam of temperature 465 °C. (869 °F)  Rapid decomposition was produced by spray-mixing the hydrogen peroxide under pressure over a KMnO4 paste, which powered the fuel turbo-pumps.  Turbo-pumps were used to force the propellants into the combustion chamber under high pressure, saving weight over using pressurized gas (which would have required pressure-resistant tanks & plumbing).  The disadvantage of turbo-pumps was mechanical complexity and reliability.  The liquid oxygen was sprayed through 18 rose jets into the top end of the combustion chamber.

It had a take-off weight of 28,300 lbs., a maximum thrust of 58,500 lbs., and a top speed of almost 3,600 fps (Mach 3.6) (2,500 mph/4,000 kmph).  The thrust lasted for 65 seconds, reaching a speed of 5,047 ft/s, and an altitude of 23.3 miles.  [AD-A434 478 (1961)]  The gases were expanded to a pressure of 0.85 atm. at the point where they left the engine.  Thrust increased at height to 29 tons, since ambient air pressure drops with altitude. The V-2 was guided by four movable graphite vanes immersed in the exhaust gases.  (And it became the basis for the Redstone missile in the U.S. in 1958.)

Rocket were built of rolled 3 mm (0.012") thick steel welded closed, to save money over the cost of aluminum. There was a double wall in the nozzle for cooling  by the methanol fuel.  Combustion chamber burning temp. was 2,700 ºC.  Combustion chamber cooled internally by a thin layer of methanol fuel forced in to the combustion chamber by thin holes in four welded rings.  Due to the pressure within the combustion chamber, the alcohol was spread out and vaporized, forming a cooling barrier between the burning fuel and the inner combustion chamber  The 18 brass Injector cups had 120 holes at its top for injection of the LOX, and the alcohol fuel was injected from the side walls of the cylindrical cup through 44 nozzles.  The fuel-oxygen vapor was then forced into the combustion chamber where it burned.

The cost of a bomber plane shot down was thirty times the cost of an A-4, and no lives were lost.

The $I_{sp,}$, the Specific Impulse, is the measure of a propellant's power, and is the kg of thrust per kg of propellant consumed per second, and thus has a unit of seconds.  LOX and Liquid Hydrogen had the highest Specific Impulse of all propellants, which Tsiolkovsky had theorized.  The Specific Impulse, of LOX + methyl alcohol is approximately 235 seconds.  The Isp of LOX + gasoline in contrast, is approx. 360 seconds, making it substantially more powerful, but the Nazis used alcohol rather than gasoline due to wartime shortages of gasoline.

The A-4 had a range of only 377 km (236 miles), because of the huge weight of the HE warhead.  After being launched it reached a maximum height (apogee) of 56 miles before descending to its target.  It landed with a crash energy equivalent to the explosion of a half ton of TNT, in addition to the explosion of its 1 ton HE warhead.

As the first crude gyroscopic, inertial-guided missile, it had only a CEP of less than three miles, which meant that 50% of the missiles would strike within a circle around the target of radius three miles.  The guidance system controlled aerodynamic rudders on its four fins located at its base, along with graphite exhaust vanes ("jetevators") to channel the engines exhaust.

It was twenty times as expensive as the Nazi V-1, the first cruise missile.  V-2 development and production costs were estimated at over $16 billion (in 2012 dollars), compared to the $36 billion (in 2012 dollars) cost of the Manhattan Project.  [Karp (1996)] [Billie (2004)]  One estimate is that each V-2 cost about $320,000 in 1990 dollars.  [Karp, p. 39 (1996)]

\*　\*　\*　\*

But interest in rockets and space travel had long preceded the Nazi V-2.  Soviet theoreticians had long established the basis of the science of rockets.

U.S. rocket pioneer Robert Goddard was (erroneously) ridiculed by the "New York Times" in 1919 for saying that rockets could function in the vacuum of space, and was generally an object of derision in the 1920's.  Goddard, an American mathematics professor, had begun his work on liquid-fueled rockets in 1909, and independently came up with some of Tsiolkovsky's ideas, including the increased efficiency of multistage rockets.  His two 1914 U.S. Patents covered the essentials of rocket propulsion:  combustion chambers and nozzles, liquid propellant feed systems, and multiple-staging.

A comparison between Goddard's early rockets (though his patents remained classified in 1942) and the V-2 was such that Goddard was "stunned by the similarities". "Goddard was convinced he could have given the United States superior rockets if he'd had the kind of government support that [captured] Germans like Wernher von Braun received."

When Goddard died he held 48 patents, which were then supplemented posthumously (through the efforts of his widow, Esther Goddard) for a total of 214 patents related to rocket technology. [Doyle, p. 3-4 (1992)]

And interest in space was not just Goddard's. Science fiction author, Arthur C. Clarke, suggested in the October 1945 issue of "Wireless World" the idea of a geostationary (also called geosynchronous) communications satellite orbiting 22,300 miles above the equator, where it's speed would match the Earth's rotation, and thus appear to be in the same spot, stationary in the sky.

The idea of reconnaissance (spy) satellite for intelligence gathering no doubt came to mind with the seminal 1946 report "Preliminary Design of an Experimental World-Circling Spaceship", a now classic RAND report.

In July 1939, the first NAS (Naval Air Service) JATO (Jet-Assisted Take-Off Rocket) contract was issued to famed Hungarian aerodynamicist von Karman and Malina. The JATO provided the extra boost required to get airplanes airborne on a short run-way, as on an aircraft carrier, which were to prove very effective in the Pacific War of World War II, as American forces rolled back the Japanese Empire Pacific island by island. [Billie (2004)]

\*   \*   \*   \*

"[The Western Development Division] created a family
of ballistic missiles that used every technology that we
knew of at the time."

-- Maj. Gen. Osmond J. Ritland (1974)

But things would improve, when the U.S. and Soviets – realizing the military importance of the new weapon -- each got a portion of the V-2 scientists and equipment for their programs. The Americans got to the research and manufacturing facility, Peenemunde first, taking 150 scientists and over 100 A4's. The US got the "cream of German rocket experts and technology", the top (initially 115, later more) V-2 personnel, 4 tons of paperwork (design, report, test and production records), and "entire trainloads of missile parts and other hardware"

The Soviets got the remainder of the scientists, who were spirited to Russia. By 1947, under a team led by Soviet scientist Sergei Korolev, they had built the first operational missile, dubbed the R-1, which was basically a modified A4, though basically it was a V-2 copy, although no Germans worked on it directly.

The first Russian design of a single liquid-fueled engine fed by a single turbo-pump was a failure, due to the high pressures and through-put demands on the turbo-pump materials. Valentin Glushko, the rocket engine specialist, and assistant to Chief Designer Sergei Korolev, in 1952, decided on multiple turbo-pumps to feed the fuel to the multiple rocket engines of the successful R-7 ICBM design that launched Sputnik I, the first satellite. [Heppenheimer (2004)]

In May 1954 USSR had made the decision to build the world's first ICBM. [Bille (2004)] In 1956, Khrushchev announced, "Bombers are obsolete." Research continued, and by 1957, the Russian performed a successful launch of the R-7 ICBM, two years ahead of the U.S.

It was a two-stage liquid-fueled rocket, fueled by LOX and kerosene, capable of carrying a 3 ton warhead 8,000 miles, capable of reaching the U.S. But it was introduced to an astonished world with the launch of the Sputnik I, the first satellite. The R-7 was designated the SS-6 "Sapwood" by NATO.

But the U.S. was not far behind. Engineers were not sure if a plane (and thus missile) could tolerate breaking the speed of sound (Mach 1), or would fall apart and be destroyed by the unknown stresses of "supersonic" speed.

In October 1947, Chuck Yeager became the first pilot to fly faster than the speed of sound (Mach 1) in the Bell X-1 "rocket plane". It was constructed of a skin of high strength aluminum, and stainless steel propellant tanks of LOX and alcohol/water fuel, like the V-2. The fuel was fed by pressurized nitrogen gas which fed the fuel-oxygen mixture to a Reaction Motors, Inc. four-chambered, 6,000 lb. thrust rocket engine. The four chambers were stacked close together, and could be fired individually or in groups.

The smooth nose of the rocket plane was patterned after the shape of a .50 caliber bullet.

*　*　*　*

"Celestial mechanics is a very old, well established art."

-- Edward Hall (1989)
oral history interview

The road to the first ICBM, the liquid-fueled Atlas, was fraught with uncertainty and difficulty. There were the unproven Navaho cruise missile program booster rockets. In fact, the basic Atlas propulsion system had a design start in the Navaho missile program.

There was the development of the Atlas airframe, a major, and uncertain undertaking. The guidance system was unreliable and unacceptably intricate. And there was no proof that reentry was feasible and workable. High rate of flow fuel and liquid oxygen pumps were uncertain, vibration effects were a question mark, and the feasibility of igniting a rocket engine at altitude was uncertain. And all had to be resolved at once in a reasonable time-frame.

Thus the backup system of the more advanced Titan I was under simultaneous development. The Titan had a monocoque airframe, a different engine, a more advance guidance system, and was a true two-stage missile. [Emme (1964)]

There are six major technologies for missile development:

1.　　　propellant performance,
2.　　　propellant production and processing,
3.　　　case material and nozzle design,
4.　　　thrust vector control (TVC), controlling the orientation of the rocket in powered flight
5.　　　navigation and guidance, which controls TVC
6.　　　re-entry vehicle design

There is also ignition, and thrust termination (TT).

A rocket is a fundamental example of Newton's Third Law of Motion, that every reaction has an equal and opposite reaction. Thus, when the high-speed exhaust from the burning fuel of the rocket, is directed out the rocket's nozzle(s) at its base, the rocket is propelled in the opposite direction.

The rocket's speed is determined by the weight of the exhaust gas, and the square of its velocity out of the nozzle – which is determined by the energy released in its combustion. Also a factor is the mass of the rocket structure, which should be minimized.

The first led to the improvement of the V-2 by changing the fuel from (LOX + ethyl alcohol) to (LOX + RP-1 (kerosene)).

The second improvement of the V-2 was to save weight by removing the double wall arrangement incorporating integral rocket casing/fuel tanks, rather than the separate fuel and oxidizer tanks inside the outer skin/superstructure of the A-4.

And there was the concept of multiple staging. As the fuel burned until it was empty in the first stage, the empty casing dropped off, reducing the dead weight for the remaining rocket to carry. Three stage rockets were the maximum developed.

Another change was to have the warhead/RV separate from the rest of the missile, which fell away. The entire V-2 landed at its target, but you got greater range not needing to insulate the missile to resist the heat of atmospheric reentry. This was similar to having a separating nose cone/warhead from the rocket. [AD-A439 957 (1990)]

Gimbaled (swiveling) engines to control the direction of flight, replacing the movable vanes in the exhaust that the Germans had used. The vanes reduced thrust by about 17%. [AD-A440 094 (1997)]

*   *   *   *

The perfection of the modern nuclear ICBM/SLBM was made possible by the perfection and optimization of the "deadly triumvirate" of the ballistic missile: a light weight, "baby" thermonuclear warhead + the accuracy-determining missile inertial guidance system + an improved missile weight and performance with lighter rocket casings and higher performance solid fuel rocket propellant.

In other words, the technological nexus for ICBM range and effectiveness in destroying its target was accomplished by:

1. minimizing the mass of the TN warhead (and additionally, the mass of the MIRV bus and inertial navigation system (decreasing the missile's required payload "throw-weight"),

2. minimizing the rocket casing mass and maximizing the percentage of rocket fuel, and

3. maximizing the accuracy of the guidance system that determined how close the ICBM hit its target, which enabled a reduced nuclear yield, and thus allowing the reduction of the yield, and thus the mass of the warhead.

The design of an ICBM (and other missiles) encompasses five major areas. There is the propulsion system, the rocket structure and aerodynamicist, and navigation guidance and control. There is the weapon system design (warhead and nose cone/reentry vehicle). And finally there is flight test, its associated radio-transmitted telemetry data, and the missile's associated land base test instrumentation.

Performance of Standard Rocket Propellants

| Oxidizer | Fuel | Storable | $T_c$ | $I_{sp}$ | |
|---|---|---|---|---|---|
| Liquid | | | | | |
| LOX | liquid H2 | No | 4180 | 450 | Space Shuttle |

| | | | | | |
|---|---|---|---|---|---|
| LOX | RP-1 | No | 5650 | 266 | Atlas |
| Nitrogen Tetroxide (N2O4) | UDMH | Yes | 5390 | 256 | Titan II |
| **Solid** | | | | | |
| AP | CTPB+HMX | Yes | | 250 | MM III |
| double-base (NC+NG) | | Yes | 1400 | 220 | Sprint |

[Karp, p.102 (1996)]


## The Evolution of the Strategic Missile

The evolution of the Strategic Ballistic Missile encompassed:

1. German V2 (1944): The baseline of strategic missile development; a rocket body with liquid oxidizer and fuel tanks inside rocket body.

2. Integrated rocket body/fuel tanks (mid-1950's): Saves weight, thus increasing range, by eliminating separate rocket outer metal casing, and metal fuel and oxidizer tank casings.

3. Atlas ICBM (1959): Cryogenic liquid-fueled ICBM. The Atlas used the increased energy of liquid propellants and saved weight with a dime-thick stainless steel skin "balloon" inflated by the internal pressure from the LOX (Liquid Oxygen) and liquid fuel propellants. Each silo cost $3.6 million.

4. Polaris A-1 SLBM (1960): The first solid propellant-fueled, strategic ballistic missile (a two-stage SLBM). Used an AP-based (ammonium perchlorate), with aluminum powder solid propellant.

5. Minuteman I ICBM (1962): The first solid propellant ICBM. An AP-based solid fuel missile that made it low maintenance, long lasting, and instantly ready to be fired, unlike liquid fuels. Also, the silos cost a much reduced half million dollars each, over those for the Atlas liquid propellant ICBM. [Bair (1988)]

6. Minuteman I ICBM/Polaris A-2 SLBM (1962): Changed missile casing from steel to a fiberglass-epoxy laminate, for weight savings.

7. Polaris A-3 SLBM (1964): Back to the balloon: casings for solid fuel ICBM/SLBM's. Kevlar-49-epoxy, then graphite fiber-epoxy (1971; Poseidon SLBM) high tensile strength fiber casing; for propellant pressure requirements, as well as increased weight savings.

8. Polaris A-3 SLBM (1964): CMDB solid propellant. Composite Modified Double-Base (HMX, NC, and NG; actually a Triple-Base propellant) solid propellant.

9. Minuteman II (1965): CTPB solid propellant binder. Carboxyl-Terminated PolyButadiene binder.

10. Minuteman III (1970): HTPB (military grade designation: R45M) solid propellant binder. Hydroxyl-Terminated PolyButadiene binder. Cheaper, easier to work with, lower viscosity, and better curing and aging characteristics than the CTPB binder it replaced. [AD-A387 318 (1995)]

11. Trident II, and Improved Minuteman III (1977): High-Energy, Cross-Linked solid propellants.

## Short-Range Strategic Ballistic Missiles

The first step on the road to Inter-Continental Ballistic Missiles was the short-range strategic ballistic missile, stationed and launched from NATO European allies, close or on the border with Russia and its East European allies/satellites. The Russians reciprocated in 1962, leading to American insecurities and the resulting Cuban Missile Crisis.

Honest John (1953)

Mk 7 initially, upgraded to W-31 30" dia. boosted Oralloy fission warhead; 1,000 or 1,500 lb. payload

Mobile, solid-fueled, unguided, spin-stabilized SRBM (Short Range Ballistic Missile)
Dimensions (missile body):  28" x 327"
Fuel:  NC + NG + Triacetin
Fuel Liner:  Cellulose Acetate
Weight (fueled, with warhead):  3,750 lbs.
Weight (burnt propellant):  2,000 lbs.
Speed:  Mach 2.4
Range:  38 km [LA-4350-MS (1969)]; 25,000 yds; about same range as 280 mm artillery shell ["GACAEC", 35th, p. 22 (1953)]; 31,000 yds. range [CIA (1978)]
Guidance:  Unguided
CEP:  1,000 ft. [LA-4350-MS (1969)]

200 deployed in Europe:  Belgium (24), Denmark (12), France (72), Greece (24), Italy (24), Netherlands (24), Turkey (36), U.K. (48), and West Germany (132)
Use:  battlefield support [Gervasi (1972)]

Corporal (MGM-5B) (1954-1966)

W-7 30" dia. atomic warhead

Surface-to-surface SRBM missile

Nose shape:  65.5" long/high
Dimension:  30" dia. x 44' high
Fin Span:  7'; delta fin configuration
Range:  70 - 80 nm
Apogee:  31 miles
Thrust:  20,000 lbs. for up to 64 seconds, at 15,000 ft. altitude, operating at a chamber pressure of 300 psi.  Regenerative, axially cooled 650 lb. (replaced by a 125 lb.) motor.
Takeoff Weight:  11,250 lbs. with a warhead weight of 1,500 lbs.
Empty Weight:  2,961 lbs.
Velocity:  Mach 3.8
Fuel:  Pumped by compressed air (initially stored at 2,350 psi), 2,100 lbs. fuel (80% aniline + 20% furfural alcohol) + 4,370 lbs. oxidizer (IRFNA, with a nominal content of 14% nitrogen tetroxide ($N_2O_4$)); hypergolic mixture; In 1948, determined minimum ignition delay with (60% furfural alcohol + 40% aniline) with IRFNA; finally changed to (46.5% furfural alcohol + 46.5% aniline + 7% hydrazine).  The IRFNA

(Inhibited Red Fuming Nitric Acid) had 14% $N_2O_4$ (nitrogen tetroxide), and was stabilized with 2.5% water, and .6% HF.

CEP: 350 m

Guidance: A quick-shutoff valve, capable of operating in a few milliseconds was installed, in lieu of the throttling back of the thrust, as had been done with the German V-2. Stability and ease of control was enhanced by replacing trapezoidal with delta-shaped fins. Gyroscopes, with accelerometers were also added to improve accuracy. In the missile exhaust, there were graphite jet vanes with a leading edge of molybdenum. [AD-A586 733 (1961)]

Contractor: Firestone Tire & Rubber

358 produced for stationing in Europe by June 1956. 1,000 produced. Of dubious reliability (25-60% successful use). Phased out starting in 1964, superceded by the Sergeant missile. Development had begun in 1944. [Clark (1972)] [Hunley (2007)] [AD434 478 (1961)]

Redstone (PGM-11) (1958 – 1964)

W-39 3.8 MT warhead; payload (throw-weight) 6,305 lbs. (1958 Operation Fishbowl Events Teak and Orange)

Single stage, liquid-fueled

Dimensions: 70" (thrust unit/fins) dia./64" (body) dia. x 69' 4" high
Weight: 16,275 lbs. (empty); 63,200 lbs. (fueled); Lift Off Weight: 61,185 lbs.
Propellant: LOX (25,090 lbs.) + Ethyl Alcohol (+ water) (18,800 lbs.)
Thrust: 78,000 lbs.
Range: 50-175 nm
Maximum Altitude: 57 statute miles
Speed (Boost Cutoff/maximum): Mach 4.8; Speed (Reentry/maximum): Mach 5.5; Speed (Impact/maximum): Mach 2.3
Time (Seconds; from Launch to Impact): 375.1 seconds
     Payload Separation: 135 s.
     Zenith: 227 s.
     Reentry: 348.6 s.
     Impact: 375.1 s.
Guidance: Inertial; two accelerometers; one with its measuring axis in the lateral direction, the other measures the range. The accelerometers are connected to a computer that determine the velocity and distance the missile has deviated from its theoretical trajectory. [AD-A995 454 (1959)] [Elliott (1959)] Carbon vanes at base of nozzle.
CEP: 660 ft [LA-4350-MS (1969)]; 300 m [Bullard (1965)] [AD-A995 454 (1959)]; 1 km [Mackenzie, p. 64, fn100 (1990)]

Contractor: Chrysler Corp. for the U.S. Army. Rocketdyne: propulsion system [Bullard (1965)] [AD-A995 454 (1959)]

120 produced. An advanced direct descendant of the V-2. Developed at the Redstone Arsenal, Huntsville, Alabama, where the 118 German rockets scientists spirited out of Germany immediately after the end of WW2 were placed.

Developed about 50% more thrust than the V-2's 56,000 lbs. at about the same engine weight. Redstone had a double-walled thrust chamber made of welded steel, similar in construction to that of the V-2, and cooled by circulation of the fuel. The fuel injectors were made of high-nickel steel alloy (maraging steel). Four large air rudders carried on fixed fins of the tail, operated in conjunction with carbon jet vanes, which projected into the rocket engine exhaust stream, to provide both path and attitude control.

Was phased out in 1964, replaced by the Pershing I.

Further evolution the engines for the Navaho booster, made by Rocketdyne, then further perfected to power the Thor and Jupiter IRBM's, and the Atlas ICBM. Their thrust chambers were made of as many as 360 formed seamless nickel tubes, which were clustered, reinforced with bands, and brazed together. A high-temperature gas system was developed for driving the turbines for the pumps. The engines were further evolved and perfected for the Thor, Jupiter, and the Atlas missile, for which there were three thrust chambers, one for each engine.

[AD830 267, p. 72 (1963)] [AD-A434 326 (1958)] [Howard-White (1963)] [Hunley (2007)]

Sergeant (MGM-29A) (1958 – 1973)

W-52; 2 warhead sections with separate yields; 225 or 512 kt possibly

Surface-to-surface SRBM that replaced the Corporal missile.

Dimensions: 31" x 34.5' (79 cm x 10.5 m)
Casing: AISI 4130 steel (tensile strength 180 ksi) motor case 0.109" thick; weighing 630 lbs.; roll and weld fabrication
Launch Weight: 10,100 lbs. (4,600 kg)
Payload: 1,600 lbs. (680 kg)
Fuel Weight: 7,069 lbs.
Thrust: 45,000 lbs. (20,400 kg)
Range: 139 km (75 nm)
Fins: AZ91-T6 magnesium alloy castings; Jet vanes: glass fiber-phenolic resin composite with molybdenum (AMS 7805) leading edges
Thiokol M-100 solid-fueled rocket motor: 63% Ammonium Perchlorate + 33% LP-33 Liquid Polymer fuel and binder (Polysulfide rubber); Five Point Star grain configuration
Nozzle: nozzle body: 1020 steel; nozzle throat insert, with high thermal conductivity in the a-b plane (horizontal; the x-y plane), high erosion resistance, and good oxidation resistance: pyrolytic graphite
Casing Yield Strength: 135,000 psi; Tensile strength: 165,000 – 195,000 psi
Pyrogen igniter; Air transportable;
CEP: 830 ft. (253 m)
Guidance: Inertial guidance system; Univac

The Sergeant was the first large missile to employ a thiokol rubber-bound composite propellant. The Sergeant was a single stage, surface-to-surface missile. The Sergeant was powered by a Thiokol XM100 polysulphide rubber composite fueled motor. This could produce 200 kN of thrust for 34 seconds.

A total of 473 Sergeants were built. [Contradiction: there were "several thousand" of Honest John and Sergeant missiles.] [JCAE, Pt. 1, p.9 (1973)] The Sergeant, together with the Honest John, were replaced by approximately 200 Lance missiles, starting in 1973 until completion in 1977. ["History of the Custody and Deployment of Nuclear Weapons," p. 132 (1973)]

Contractor: Sperry Rand for U.S. Army
[LA-4530-MS (1969)] [Cagle (1971)] [Hunley (2007)]

Lacrosse (MGM-18) (1959 — 1964)

W-40 warhead; Weight:  583 lbs.

Single-stage, solid-fueled GLCM (surface-to-surface)

Dimensions:  20.4" dia. x 19' 2" long; Wingspan:  9'
Weight:  2,360 lbs. at launch, 1,645 lbs. at burn-out
Propulsion:  Thiokol solid-fuel rocket
Range:  12 - 20 miles
Altitude:  5,000' or 15,000'
Speed:  Mach 2; also Mach 0.8; also 1,500 ft/s
Guidance:  Radar-guided

Main Contractor:  Martin Marietta for the U.S. Army; Contractor (Propulsion):  Thiokol


Little John (M-51) (1960 – 1969)

W-45 warhead

Single-stage, solid-fueled
Propellant:  NC + NG + 2-Nitrodiphenylamine
Rocket Liner:  Pyrolock
Dimensions:  12.5" dia. x 14.5' long
Span:  30"
Weight:  980 lbs.
Speed:  Mach 1.5
Range:  11.3 miles
Guidance:  Unguided

Contractor (Propulsion):  Allegany Ballistics
Main Contractor:  Emerson Electric


Pershing I (1964) & Pershing IA (MGM-31) (1969)

W-50 warhead; warhead sections choosable with 3 different yields (400 kt/200 kt/60 kt)

Possible RV:  34.1" base dia. x 51" high; 14° cone half-angle; hemispherical tip (0.159 m/6.25" radius) on top of conical RV; weapon storage volume in RV:  34.1"/0.87 m base dia. x 39.4"/1 m high; the RV's initial ballistic coefficient, $W/C_D A$, is approximately 537 kg/m2 (110 lbs./ft2)

Nominal reentry conditions are 6.1 km/s (20,000 ft/s) at an angle of 23° below the local horizon; the peak heating rate is on the nose tip, and is 181 cal/cm$^2$.s for about 20 seconds); 45.5 lbs. total weight of the lap-wound fiber glass-phenolic flat tape RV outer insulation/ablation layer. The ablation layer was over a 0.054" thick aluminum substructure.  [AD465 896 (1965)]

2-stage, inertially-guided, Thiokol solid-fueled, surface-to-surface IRBM; "terminally guided" RV using radar to compare stored terrain image of target.

Size:  Pershing first stage:  34.5' high x 40" dia.; Pershing 2$^{nd}$ stage:  8' high x 40" dia.
Propellant:  4,451 lbs. in the first stage; 2,875 lbs. in the second stage  [TM 9-3305 (1981)]
Missile External Casing Thickness:  First stage: 0.090"; Second stage:  0.068"

Tensile strength of case material:  195,000 to 22,000 psi  [AD818 416 (1967)]
Case Material:  D6AC steel (220 ksi tensile strength)
Weight:  10,150 lbs. (4,600 kg)
Range:  Pershing 1A:  740 km/465 miles [LA-4350 (1969)]; 480 m/725 km [JCAE, Pt. 1, p. 9 (1973)] [AD-A017 242 (1975)] [CIA (1978)]
Speed:  Mach 8 at burn-out
CEP:  0.25 nm

When the first stage rocket motor is ignited, Pershing 1A liftoff occurs and the missile begins a predetermined pitchover maneuver toward the target.  When first stage burnout is achieved the missile enters a coast period.  At the end of the coast period, first stage separation occurs and the second stage motor is ignited to accelerate the remaining missile sections along the flight path.  During the second stage burn, the guidance computer constantly computes the missile's velocity and displacement.  When the proper values of altitude, range, and velocity have been attained a thrust termination signal is applied, second stage separation occurs, and the RV/warhead is spun at 32 rpm for aerodynamic stabilization of its forward motion as it continues on a ballistic trajectory to the target. [AD-A017 242 (1975)]

The first and second stage contain a solid propellant grain, with a circular hole running lengthwise throught the grain.   The propellant type used is a PBAA composite (AP + Al + hydrocarbon). [AD818 416 (1967)]  The Pershing I Thiokol first stage engine produced 26,290 lbs. of thrust for 39 seconds.

Designed to replace the Redstone IRBM in Europe.   126 launchers stationed in West Germany. [JCAE, Part 1, p. 9 (1973)]  [Talbott (1984)]; 330 missiles [LA-4350 (1969)].


Lance (MGM-52) (1972 – 1992)

Originally carried the W-70, but was upgraded to an ER (Enhanced Radiation – the Neutron bomb) with the W-70Mod3 (ER) in 1981; 469 lb. (non-ERW W-70) nuclear payload [AD-A017 242 (1975)]

An excellent replacement (surface-to-surface IRBM) for the obsolete Sergeant and Honest John missiles.  The Lance missile is made primarily of aluminum and has large control surfaces (fins).  The Lance missile motor case is made primarily of aluminum alloy (2024-T6), with a bell-shaped thrust chamber. [AD818 416 (1967)]  All-weather, day or night, long range rocket.  Cheap, mass-produced, low maintenance, mobile, storable bi-propellant, liquid-fueled rocket.  Air transportable, including by helicopter.  Road mobile (30 mph) as well as off-road capability.

The last of around 1,000 Lances [AD-A017 242 (1975)] were withdrawn from European service in 1992.

Dimensions:  21.2" dia. x 20' 6" long   [BC73-60 (1973)]
Total Weight:  2,844 lbs. with nuclear warhead
Storable Liquid Propellant:  Bottom tank contains the oxidizer IRFNA (Inhibited Red Fuming Nitric Acid, with 2-3% water, 14% nitrogen tetroxide (N2O4), as stabilizers, and 0.6% HF (hydrogen fluoride) as an anti-corrosive), and fuel is UDMH (Unsymmetrical Di-Methyl Hydrazine).  Fuel inlet is located at engine axis.
Engine:  173 lb. weight.  19.4" long by 21.1" diameter.  Starts by rupturing of burst discs for propellant and oxidizer; casing (including motor casing) case material:  aluminum alloy 2014-T6
Case Material:  Yield Strength:  59,000 psi transverse, 60,000 longitudinal; Tensile Strength:  67,000 transverse, 68,000 psi longitudinal
Nose Cap Material:  The Lance system utilizes skins of aluminum-phenolic-paper honeycomb and bonded cork-silicone ablative shielding. [AD787 040 (1974)]
Boost Phase:  During the boost phase the missile accelerates under full thrust until after 1 to 7 seconds  it attains the velocity needed to carry it to the target; 50,000 lb. class; annular bell nozzle booster; forged steel thrust chamber body with  ablative liner; cast aluminum injector; during boost phase (6.1 s), both boost and sustainer thrust chambers operate; Engine mixture ratio:  3.4:1

Sustainer Phase:  5,000 lb. class; center bell-nozzle sustainer (throttleable); sustainer can be turned on and off as required.  Sustainer time length (114 s) and ratio (3.19:1)

Nozzle Area Ratio:  Booster:  5.7:1; Sustainer 4:1

Chamber Pressure:  ~950 lbs. (booster and sustainer)

TVC:  Side force of 400 lbs. from any of 4 separate boost phase TVC valves with a fuel flow of 5.5 lbs./s at 950 psig inlet pressure.

Range:  115 km maximum [AD-A127 320 (1983)] [AD-A206 251 (1988)]; 5 to 125 km [JCAE, Part 1, p. 10 (1973)]; 100 miles (160 km) [CIA (1978)]; 120 s maximum mission duration.

Thrust:  47,000 lbs.

Accuracy:  375 m CEP at 125 km

Manufacturer:  Ling-Temco-Vought (LTV)

555 In Service:  Belgium (60), Sicily, Italy (60), Netherlands (60), U.K. (200), and West Germany (175).

The guidance set consists of directional control (with the RSG-15 gyroscope as primary sensing device), and velocity control electronics (the primary sensing device is an accelerometer).  [TM 9-3305 (1981)]

The rocket engine was developed by Rocketdyne jointly under the U.S. Army Missile Command, and LTV Aerospace System Division.  Propellants pressure-fed at approximately 1,175 psi (by ignition of a solid propellant gas generator) from their propellant tanks, and the Lance is designed to last five years in the field.

Gas generator also provides a gas flow to side-mounted spin nozzles for missile rotation for stabilization.


## Strategic IRBM's


Helmut Grotttup was an electronics engineer, who deliberately surrendered to the Soviets along with his team.  Lightweight copper motors were one improvement of the first ICBM, the Soviet R-7. [Ford (2011)]

Air Force General Bernard Schriever, as the first head of the Air Force Air R&D Command's Western Development Division, appointed in 1954.  He shepherded the development of the Thor IRBM, the Atlas, and the Titan liquid-fueled ICBM's, then moved on to the Minuteman solid-fueled ICBM program.

The most powerful rocket fuels are liquids -- they have the highest energy per unit of mass and volume.  They powered the first ICBM:  the Atlas, one of which sits outside in front of the Science and Technology Museum in Ottawa, Canada.  It was developed in 1959, and fueled by liquid oxygen and kerosene.  The Saturn V booster for the civilian Apollo moon missions of the late 1960's and early 1970's, and the Space Shuttle were also liquid-fueled, the Shuttle by LOX and liquid hydrogen, the most powerful liquid fuel of all, and both cryogenic liquids.

The earliest strategic IRBM's took 15 to 20 minutes to reach ther target, and were launched to reach a maximum height (apogee) of a few hundred miles above the earth's surface, releasing an RV that descended on its target at around 7,000 - 8,000 mph, and lacked accuracy and warhead weight.

The ICBM was developed soon after, and traveling a longer distance took about 30 minutes to reach its target.  It had an apogee of 750 - 900 miles, and released an RV whose maximum speed attained 15,000 mph.

Thor IRBM (SM-75, later the PGM-17) (1959 – 1963)

1.44 MT W-49, weighing 1,600 lbs. RV: First the GE Mk 1, then the GE Mk 2 (beryllium/copper heat sink) RV; 5' high x 5.1' dia.; 1,375 lbs./624 kg empty (without the warhead) RV, then the GE Mk 3 (128" x 32" x 44")

Single Stage liquid-fueled IRBM; its operation was very unreliable and undependable.

Fuel: LOX + JP-4 ($C_{9.5}H_{19}$) kersosene (aircraft turbine engine fuel)
Dimensions: 96" dia. x 64.8' high
Dry Missile Weight: 7,598 lbs.
Takeoff Weight without payload: 104,238 lbs.
Single gimbal-mounted engine, 165,000 lbs. thrust, 146.4 second burning time
Specific Impulse: 250 sec
Propellant Weight: 96,940 lbs.
Propulsion System: Rocketdyne LR79 liquid-fuel booster, with Rocketdyne LR101 vernier liquid-fuel rockets.
Range: 1,500 nm (2,900 km); Vertical range: capable of lifting 16,000 lbs. to 300 km; capable of lifting 5 MT device to 105 km.
Apogee: about 350 miles high with a 7,000-8,000 mph reentry speed.
CEP: < 2 nm
Contractor: Lockheed/Douglas

Deployed: 60 in England 1959-1963; from November 1962 to August 1963 all were retired; Based on Atlas ICBM technology. Cost: Thor booster and launch is $1.7 million in 1961 dollars ($15 million in 2012 dollars)

[Mackenzie p. 120-121,131n110 (1990)] [Greenwood, p. 164 (1975)] [Baker (1978)] [Stine, p. 198 (1991)] [NRL Report 5097 (1957)] [Goldberg, part 2, p. 875 (1981)] [Yenne, p. 52-53 (2012)]


Jupiter IRBM (PGM-19) (July 1960 – 1963)

1.44 MT W-49 weighing 1,600 lbs.

RV (alone): Ablative, 9' high x 65" base dia., 2,715 lbs. (Contractor: Goodyear) [SCDR 99-59 (1959)] [Heppenheimer (1997)] [Greenwood (1975)]; Jupiter nose cone has a hemispheric top, followed by a cone-shaped 13.3° semi-vertex angle [AD158 516 (1957)]; RV spin: 1 rotation/s; Nose tip ablation: less than 0.375" around the nose tip base, and less than 0.2" on the extreme tip.

Liquid-fueled, single stage missile

Fuel: LOX + RF-1 (kerosene)
Dimensions: 105" dia. x 60' high
Single gimbal-mounted engine swiveled by hydraulic actuators; with the small 1,000 lb. thrust swivel-mounted vernier rockets. Baffles installed in LOX and fuel tanks to counteract sloshing that made the missile unstable.

Range: 1,540 nm (2,845 km)
Missile dry weight: 10,715 lbs.
LOX            67,645
RF-1 (kerosene)   30,209
Total weight:    108,804
Propulsion System: Rocketdyne JR79 liquid-fuel rocket
Thrust: 150,000 lbs.
Re-entry: 66.5 s (from 100 km)

Re-entry speed:  Starting from Mach 15.45 at 100 km, reduced to Mach 0.5 by impact because of aerodynamic atmospheric drag
Max. missile speed:  Mach 15.45
Total flight time:  1,016.9 seconds
Main Engine Cutoff:  158 seconds
Zenith (Max height of trajectory):  660 km (356.9 nm)
CEP:  0.5 nm  [Mackenzie, p. 131, n110 (1990)]

Contractor:  Chrysler Corp.

30 and 15 deployed in  Italy and Turkey, respectively, 1959 with retirement in April 1963.  An evolution of the Redstone SRBM.  [Grimwood (1962)] [Neufeld (1990)]

Pershing II IRBM (MGM-31) (December 1984 - 1991)

W-85 Oralloy warhead, 12.4" dia. x 41.7" long; throw-weight:  700 lbs. (RV + W-85)

Two-stage solid-fueled

Dimensions:  3.3' dia. x 34.5' high; second stage motor case 40" in diameter [AD268 330 (1961)]
Weight:  16,450 lbs.
Speed:  Mach 8
Altitude:  150 miles
Aerodynamics:  4 small fins at base of RV
Rocket Casing:  Kevlar-49 fiber/epoxy (Kevlar:  370 ksi tensile strength)
Range:  almost four times the range of the Pershing IA; 1,500 mile/2,400 km range, making it a strategic missile capable of hitting Moscow from West Germany
Guidance: A new automatic gyrocompassing system [AD-A067 516 (1978)]
Accuracy:  Highly accurate; 20-40 m CEP

180 based in West Germany; 54 in the U.K.

The Pershing II's very high accuracy permitted the use of a smaller nuclear warhead than were required with older, less efficient systems.  [AD-A148 828 (1984)]  The second stage motor case is 40" in diameter.

Maraging steels are ultrahigh strength steel, unusual in that they have very low carbon concentration.  They substitute elements such as 12 - 30% nickel, molybdenum, and titanium to achieve an age-hardened martensitic structure, rather than the usual quench and temper heat treatment.  Maraging steels possess one of the highest combinations of strength, toughness, and fracture resistance of any commercially available alloy.

Maraging were selected for large boosters also because of their ease of forming in the annealed condition, the relatively simple aging treatment for developing comparatively high strength levels in thick sections, good weldability, and good toughness.  The work-hardening coefficient of the alloy is lower than that of other alloy steels.  Thus deep drawing and shear spinning are feasible process for forming cylindrical shapes.  There is a tendency for designers to select the 300 grade (280,000 – 290,000 psi yield strength, or higher) [AD818 416 (1967)]

Maraging steels contain a high concentration of nickel (18% for Army missile motor cases) which ensures a complete transformation to martensite even with a very slow cool from the austentization temperature.  There is an increase in the thermal hysteresis between the formation of martensite upon cooling and austensite reversion on heating which allows the aging of the martensite matrix at elevated temperatures (850 - 950 °F).  The precipitation reactions that occur upon aging the martensitic matrix are mainly responsible for the ultrahigh strength, and hence the term "mar[tensitic] aging".

The motor maraging steel contains 18% nickel and between 8 - 15% cobalt.  [AD-A180 927 (1987)]


## Liquid-fueled ICBMs


Atlas D/E/F (September 1959 – November 1960; December 1959 – January 1962; March 1962 – December 1962, respectively) (SM-65)

First U.S. ICBM.  Unreliable and prone to failure [Neufeld (1990)].

Warhead:  Lighter 1.44 MT W-49 in Atlas D/E/F, replaced the 3 MT W-38 in earlier Atlas'.

RV:  early Atlas D's used GE Mk 2 beryllium-copper heat sink RV's; later Atlas D's upgraded to GE Mk 3 ablative RV, 3 m high; Atlas E/F used GE Mk 4 RV

Atlas F throw-weight for Low-Earth Orbit:  13,200 lbs. [OTA (1988)] (which meant a greater throw-weight for ICBM use)
Fuel:  Liquid Oxygen (LOX) (18,600 gallons) + RP-1 (kerosene in the C12 region with an H/C ratio of between 1.95 and 2.00) (11,500 gallons)
Fuel Loading:  Performed just before launching, the RP-1 was loaded first, in a time of 4 minutes and 45 seconds, followed by the loading of 5,000 gallons of Liquid LOX per minute at -500 ºF under 6,000 psi, in a time of 4 minutes and 50 seconds.  [Neufeld, p. 202-203 (1990)]
Range:  5,500 nm (10,500 km) [Goldberg, part 2, p. 872 (1981)]; Atlas D:  Range of 8,700 nm, with 2,000 lb. payload (W-49 warhead) [AD851 310 (1960)]
Dimensions:  10' diameter x 82.5 ft. high
Weight:  267,136 lbs. (fueled) (Atlas E); booster jettison rate of 7,197 lbs. (Atlas D).
Specific Impulse:  251 and 219.3 seconds, booster and sustainer engines, respectively, for Atlas D.
Thrust:  296,400 lbs. total from two LR89 liquid fuel boosters, and a 136,000 thrust sustainer engine with two LR105 1000 lb. thrust vernier engines; all engines made by Rocketdyne.
Speed:  Mach 22 (Atlas E)
Apogee:  750-900 miles above the Earth, descending at 15,000 mph
Guidance:  Atlas E:  First missile with all-Inertial Guidance System; the previous Atlas D had used radio control + Inertial Guidance.
CEP:  Atlas D:  estimated at 2.5 nm; with a 2,000 lb. payload (W-49) and resulting 8,700 nm range, 2 nm
CEP:  Atlas E/F:  1.5 nm.

Contractor:  Convair Division of General Dynamics
Deployed:  California, Wyoming, Nebraska, Texas, Kansas, New York, Washington, and New Mexico. 1961-1964 maximum deployment of 99; by June 1965 all had been retired

To reduce weight, had integral fuel tanks/rocket casing of thin, dime-thick stainless steel "balloons", pressurized by LOX and fuel.  Development started in May 1954, with successful results of 1954 Operation Castle.  It was given a head-start with the rocket engines and guidance system developed for the Navaho Cruise Missile.  It used inertial guidance.

The stainless steel skin was 0.020 - 0.040" thick, depending on loads, and kept pressurized at 10 psi.  A single insulated bulkhead separated the LOX and fuel tanks.  A special stainless steel was used because of the cryogenic LOX, and special techniques were used to weld the thin stainless steel sheets. [Stine, p. 194-196 (1991)] [Heppenheimer, p. 182 (2004)] ["History of Jupiter Missile System", p.4] ["SAC" missiles web homepage] [Poirier, p. 95 (2000)] [Greenwood, p. 164 (1975)] [Goldberg, part 2, p. 872 (1981)]

Titan I (SM-68) (April 1960 – total phase out in 1965)

Warhead:  1.44 MT W-49 weighing 1,600 lbs., upgraded to 3 MT W-38; weighing 3,000 lbs. [Shelton (1988 or 1995)] [Francis, p. 135-136 (1995)] [Stumpf, p. 31, 36 (2000)] ["1962 Pacific Nuclear Tests" (1964)]

RV:  AVCO Mk 4 ablative  [AD-A439 957 (1990)], spin fin-stabilized; weight:  1,000 lbs.  ["1962 Pacific Nuclear Tests" (1964), p. L-C-6-2, L-C-7-3]; RV is 2' 9" high.  [T.O. 21-SM68-1 (1964)]

First true multi-stage ICBM; a two-stage liquid-fueled ICBM; first 2-stage missile.  More reliable than its Atlas predecessor, but still had reliability problems of its own.

Dimensions:  10' maximum dia. x 98' high.
Weight:  110 ton fully loaded with fuel
Range:  5,500 nm (also quoted as 8,400 nm)
Apogee:  700 statute miles
Maximum Speed:  15,000 mph
Fuel:  LOX + RP-1 ("Rocket Propellant-1", kerosene in the C12 region with an H/C ratio of between 1.95 and 2.00); helium gas storage spheres pressurize the RP-1 fuel tanks and LOX tanks to 3,100 psi.; fuel tank below the liquid oxygen tank in both stages.
First Stage:  10' high x 10' in diameter; 7,750 gallons of fuel + 12,400 gallons LOX; 300,000 lb.f thrust from two 150,000 lb.f each of the two LR87 liquid-fueled, thrust booster engines by Aerojet-General.  $I_{sp}$ =  249 sec both at sea level;  In a vacuum: 80,000 lbf total thrust,  and  $I_{sp}$ = 311 sec.
Staging Rockets:  Two mounted 180° apart on the outside of the bottom of the Second Stage engine compartment.  At first stage separation (by the release of 4 staging separation bolts), they provide 9,600 lbs. of thrust for approx. 3 seconds, producing a minimum separation distance of 10' between the first and second stages.
Second Stage (fires in the vacuum of space):  8' high x 8' in diameter; 2,027 gallons of fuel + 2,985 gallons LOX; single 80,000 lb.f. LR91 liquid-fueled, thrust sustainer engine by Aerojet-General Corp.  $I_{sp}$ = 311 sec.  The total operating time for both stage's engines approaches 6 minutes for maximum range.
Vernier thrust:  900 lbs. at 250,000' altitude; there are 4 vernier nozzles spaced 90° apart around the aft end of the Second Stage.
Missile Construction:  Self-supporting, rigid framework of a copper-aluminum alloy, the thickness of a half dollar.  Stainless steel (10% nickel) seamless tubes were used in the 2 thrust chambers made by Aerojet-General Corp.
CEP:  1.0 nm

Contractor:   Glenn L. Martin (and Martin Marietta)

Like the Atlas ICBM, the Titan I's walls of the fuel and LOX tanks also serve as the skin for the missile.  First stage had at each quadrant 4 large longitudinal braces, called longerons, attached by spot welding.  Not only took thrust loads for the engines, but provided mounting points for the giant missile to sit on its launch stand.  Avoided drilling holes in the fuel tanks that could leak.  Anticipated gravity forces a maximum of 6 g's.  More sophisticated and powerful than Atlas ICBM. [Howard-White (1963)]

Titan II (April 1963)

Mounted with 8.9 MT W-53, weighing 6,200 lbs.  Later could also carry 20 lighter weight RV's (eg. Polaris A-3 MRV's). [Goldberg, Part B, p. 546 (1981)]

RV:  ablative sphere-cone GE Mk 6; nose half-angle:  about 13°; 57.3" base dia. x 122.3" high  [Hansen VIII-352 (1994)]

Two-stage, storable liquid-fueled ICBM. Only 54 missiles in Service over many years, and not retired because of its large payload. [Goldberg, Part B, p. 788 (1981)] Throw-weight for LEO (Low Earth Orbit): 7,900 lbs. [OTA (1988)]

First Stage: Driven by 2 x 215,000 lb. thrust motors fueled by Aerozine-50 (50/50 by weight hydrazine ($N_2H_4$) - UDMH mixture) fuel + nitrogen tetroxide ($N_2O_4$) oxidizer; 15,000 gallons of fuel + 17,000 gallons of oxidizer; hypergolic.

The nitrogen tetroxide is mixed with 25% nitric oxide (NO) gas to lower the freezing point to -65 °F. Could be stored in mild steel tanks without corrosion, as long as anhydrous. The chemical energy released when 1.02 gram.moles of N2O4 reacts with 1 gram.mole of Aerozine-50 (0.6522 moles hydrazine + 0.3478 moles UDMH) is $1.54 \times 10^5$ calories. [AD-A149 597 (1984)]

Combustion chamber pressure approximately 3,000 psia. Regenerative cooling was used using the Aerozine 50 in combination with the combustion chamber having a thermal barrier coating, by liquid oxidizer film cooling. The stainless steel combustion chamber had 84 fuel injectors, and was lined with ablative material (like, for example, phenolic-resin impregnated graphite cloth). Phenolic-impregnated chopped fiberglass fabric molded inserts were used as liners, to protect the especially harsh environment of the chamber throat.

Lower than the chamber throat was the ablative exit cone, which consisted of a conical ablative liner contained in a steel shell and retained by a metal ring on the aft end. The liner contour is identical to that of the regeneratively cooled expansion nozzle, both of which have a 35° half-angle cone. The liner is fabricated from a phenolic-impregnated tape, which is wound at a 20° to the longitudinal axis of the exit cone and molded to the required nozzle contour.

An excellent gas sealant, RTV-60, a silicon rubber compound, was used at all ablative-to-ablative and ablative-metal joints in the chamber. A zinc chromate putty was used as a joint filler and sealer medium. [AD368 640 (1965)] Ideal internal pressure was 11.5 psi for both the fuel and oxidizer tanks.

Second stage (vacuum): A single engine, unlike the first stage, but using the same liquid propellants; Nitrogen tetroxide + Aerozine 50 [kerosene]

Range: **6,100 nm** ["Recommended FY68 – 72 Strategic Offensive and Defensive Forces" (1966)] / 5,500 nm [Goldberg (1981)]
Weight: 148 metric tons (300,000 lbs.)
Dimensions: 3 meters (10') in dia. x 30 meters long; 0.5 cm thick skin, explosively formed from 2040-0 aluminum [AD748 416 (1972)] Has a 10' diameter skirt at base of 0.04" thick 2014-T6 aluminum skin riveted to vertical longerons and horizontal frames. [AD-A112 526 (1973)]
Speed: 17,000 mph
Thrust: 430,000 lbf, $I_{sp}$ = 259 s. both at sea level; 100,850 lbf, $I_{sp}$ = 315 s. in a vacuum.
RV adapter/spacer, and decoys weighed 805 lbs. in additional weight (totaling 8,380 lbs.).
CEP: 1.0 nm

Contractor: Martin Marietta

["Life" magazine (November 1987)] [Stumpf, p.36, 49-66, 181-2, 254, 295-296 (2000)] [National Atomic Museum web page, circa. 2002] [Braun (1975)] [Gibson (1996)] [Clark (1972)] [Doyle (1992)] [Goldberg, part 2, p. 873 (1981)]

135 Titan II's manufactured, and armed with the 8.9 MT W53; they represented a quarter of SAC's total Megatonnage. But the peak number of Titan II's in 1967 was 63 [Yenne (2012)], then finally reduced to 54 by 1970. W-53 had air-burst and surface impact options; air-burst height approx. 16,000 feet. Titan II silo hardness: 300 psi

The 8.9 MT W-53 Titan II warhead could be replaced by a 35 MT warhead by replacing U238 secondary tamper to Oralloy "Tuba" design [RDD-7 (2001) Item V.C.2b & c]. U238 to Oy upgrade results: increase of "more than twice" yield-to-weight ratio [RDD-7 (2001), Item V.C.2c].

**Solid-Fueled ICBMs**

"The rocket's red glare, bombs bursting in air."

-- "The Star-Spangled Banner" (1814)
Francis Scott Key

The development of solid-fueled rockets was a gigantic leap forward in ICBM development, performed by Col. Edward Hall, in the late 1950's as director of the Minuteman ICBM program, working with engineers at companies such as Thiokol and Boeing.

They overcame the problems of solid fuel at the time of insufficient power, and casing burn-through, and engine shut-down on command.

To increase thrust, they began using a mixture of ammonium perchlorate (an oxidizer), and aluminum powder (the fuel). The fuel-oxidizer mixture was bonded with a rubber-like polymer as a binder that also burned, and made the solid fuel castable into the rocket casing. Its rubbery nature also kept the fuel from cracking over time, which would cause the rocket to explode upon launch.

The ammonium perchlorate (AP) was ground as fine as possible – 25 micron (μm) in diameter particles – which increased the burning rate considerably. For the upper stages of the ICBM/SLBM, it had been large grain double-base solid propellant, that could be reliably ignited and burned had been demonstrated in 1955.

For even thrust, they changed the rocket burning method, using a technique developed at a small English laboratory. Instead of using the end-burning technique then used, they cast a star-shaped void all the way through the center of the fuel, which allowed the propellant to burn from the inside out evenly. With a constant burn surface area, this provided even power output throughout the ICBM's fuel burn. The igniter was at the top of the fuel casing, at the opposite end as the rocket nozzle. The casing is built to withstand the internal pressure when the rocket is burning. The solid propellant (or grain) is usually bonded to the inner wall of the case. The nozzle channels the discharge of the combustion products , and because of its shape accelerates them to supersonic velocity. As an example, a missile with an internal pressure of 70 atmospheres (~1,000 psia), the pressure goes down to 1 atm. on exiting the nozzle, producing the thrust.

The characteristic velocity is $C = (p_0 \times A) / (dm/dt)$ in m/sec, where $p_0$ is the pressure in the combustion chamber, A is the nozzle minimum throat area ($m^2$), and $(dm/dt)$ is the mass flow rate ejected by the rocket (in kg/s). The velocity of the combustion products leaving the combustion surface is at least two orders of magnitude higher than the burning rate. [AD-A425 146 (2004)]

Up to 1961, flame temperatures were 5,500 – 6,200 °F and rocket nozzles could be designed on heat-sink principles, with the surface temperature of the nozzle below the melting point of the exposed material. For flame temperatures of 6,200 – 7,000 °F (1961 – 1965) nozzles will have to be ablative, "burning off." For flame temperatures of greater than 7,000 °F, it was necessary to have liquid-cooled nozzles. [AD-268 311 (1961)]

Increasingly severe operating environments generated by increasingly high performance motors, have since 1970, caused significant changes in solid rocket nozzles, substantially improving their functionality. The main function of solid rocket motor nozzle is to channel and control expansion of hot gases coming form the chamber, thus creating the motor thrust. In order to design a rocket nozzle, important parameters include nozzle throat diameter, exit cone expansion ratio and half-angle, and aft opening diameter of the casing. The nozzle can be external to the motor or submerged into the combustion chamber. An external configuration is usual with a fixed nozzle, while the submerged configuration is usual with movable nozzles. The older four nozzle design was used originally for its ability to provide thrust vector control.

For ICBM boosters, generalized nozzle design requirements include a burn time of 120 seconds, an erosion rate of 15 - 20 mils/second, and an expansion ratio of 10. In the early 1970's, all of the flame side parts of the nozzle are tape-wrapped phenolics, with carbon and graphite cloth phenolics the most common. The exit cone required a metal shell for structural support because of the charring of the phenolic, with additional insulative material added to prevent the metal shell from melting.

The all-ablative design is critical in terms of amount of erosion and char depth. Char refers to the phase change in the material when the temperature reaches the point when the phenolic resin pyrolyzes. Pyrolytic graphite, introduced in 1962 (with the Sergeant missile RV), but widely adopted only around 1978 for nozzle throat liners, has a greatly increased resistance to erosion compared to the ablative materials. The erosion resistance partly because of the high density of carbon atoms in pyrolytic graphite.

With regards motor casings, fiberglass/epoxy was used in the Polaris and Poseidon, and upgraded to the superior (lower density, and higher in specific strength) Kevlar fiber/epoxy in the follow-up Trident I, Pershing II, and MX/Peacekeepers. The epoxy is convention bisphenol-A-based. [AD-A199 356 (1988)]

*    *    *    *

Solid propellants evolved from ammonium perchlorate-based, to double-base, to triple-base, composite double-base (CDB) fuels. The CDB propellants were based on the most powerful military high explosive commonly in use, HMX, mixed with "double-base": nitroglycerin/nitrocellulose (NG/NC). Though all three (HMX/NG/NC) are explosives, they simply burn (vigorously!) when used as missile propellant fuels.

Solid propellant ICBM/SLBMs use one of two types of high-energy propellants.

One is a composite propellant consisting of a heterogeneous mixture of approximately 70% AP, 16% aluminum powder, and 14% polymer binder to, among other things, prevent propellant cracking due to aging. (The most modern binder is HTPB, hydroxyl-terminated polybutadiene, a hydrocarbon polyurethane, first used in the Trident I SLBM.) The propellant's Specific Impulse is about 260 seconds.

The second type of solid propellant is a very high energy, cross-linked composite, modified double-base (triple-base) composition consisting of ~52% HMX, 18% NG, 18% Aluminum powder, 4% AP, and only 8% cross-linked polymer binder (by the use of a long-chain diisocyanate that allowed an increased fuel and oxidizer percentage). Its Specific Impulse is about 270 s, giving an increased range of the missile of about 8% over the first AP-based propellant, and is used in the Trident I and II SLBMs. [Smith (1980)] [Drell and Peurifoy (1994)]

## Solid Fuel Propellants

Solid fuel missile propellants consist of two different types: double-base and composite, with double-base being more powerful, but more dangerous to handle.

Double-base propellants are based on nitrocellulose mixed with nitroglycerin as an explosive plasticizer to produce a homogenous colloid. (Eventually double-base propellants evolved into triple-base, by the additions of an HE, like HMX.)

The composite propellants consist of a solid, finely particulate oxidizer (ammonium perchlorate) dispersed thoroughly through a rubbery or resinous fuel material matrix, which also prevents stress cracking, for stability and long-term storage.


## Thiokol Rubber Fuel


In 1942, John Parsons created a rocket motor that employed asphalt as a binder and fuel component, combined with potassium perchlorate as an oxidiser. The binding agent was used to maintain the optimum shape of the combustion chamber. Ammonium perchlorate replaced potassium perchlorate as the standard oxidiser in the late 1940's. This reduced smoke production and increased specific impulse. Charles Bartley inproved on this idea by replacing the asphalt with thiokol LP-2, a polysulfide rubber. The polysulphide rubber binder gave improved performance over asphalt, particulary in terms of storage temperature range and hardness.

He also developed a star-shaped combustion channel running throughout the motor. The earlier asphalt motors had burnt propellant from end to end - the same mechanism as used by gunpowder burning rockets and fireworks. However, composite solid propellants tend to burn slower than blackpowder, generating the need to expose a greater surface area of propellant at any moment in time. In addition, the rubber binder allows the propellant to bond to the motor case, enabling a channel to exist in the middle of the rubber based fuel. However, a cyclindrical channel proved inefficient. The longer the motor burns, the greater the surface area of exposed rubber bound propellant. To keep burn rate consistent throughout the flight, a star shaped cross-section is employed. This was initially referred to as a *burning star* configuration. This core burning design also allowed for lighter motor cases, as the rubber propellant acted as an insulator to protect the casing

Further research into polysulfide rubber-perchlorate propellants resulted in the "Thunderbird" rocket. This vehicle demonstrated a polysulphide rubber composite-propellant, internal-burning star-grain motor in 1947. The vehicle had an acceleration of 100 g's.

This idea evolved into the motor for the Sergeant missile (MGM-29A).

Two engineers, Keith Rumbel and Charles Henderson, then found that adding aluminium significantly increased the specific impulse of the solid propellant. They used 21% aluminium powder, 59% ammonium perchlorate and 20% plasticised polyvinyl chloride. Thiokol has added aluminium to their rubber binders ever since.

As a composite binder, first in the 1940's there was asphalt (used with potassium perchlorate). Then there was actual rubber, either natural or synthetic. In 1950, polysulfide were found by Thiokol to be the first to be able to case-bond the AP (ammonium perchlorate) propellant to the inner surface of the casing.

An internal-burning, star-shaped cavity was used. As it burned, the surface area remained more or less constant, so the thrust remained constant. It was found that the star points needed to be rounded to prevent cracking. The mandrel around which the solid propellant was cast was coated with Teflon for ease of removal. [Hunley, p. 226, 229 (2007)]

In the mid-1950's ARCO (Atlantic Research Co) discovered that 5% aluminum powder replacing 5% of the oxidizer (like AP) increased performance by 15%. There were also small aluminum staples mixed in with the fuel that further increased performance.

Then there was the search for high-energy polymeric binders, to do the double-duty of binder **and** fuel together.


"However, when we tried to produced high molecular weight polyurethanes or polyesters, we found that the normal methods did not apply. All of these processes involved heating the reaction components at elevated temperatures. (This is something that one does not do when dealing with high energy [AP] explosive compounds!). ... Fortunately, Rodney Fischer was on the alert and discovered an obscure reference in a German patent that indicated that iron chelate compounds would catalyze the reaction of alcohols with isocyanates to make urethanes at essentially room temperature. We immediately tried this procedure using Ferric Acetylacetonate (FeAA) as the catalyst, and it worked!" ["Aerojet", IV-99 (1995)]

In the early 1950's Thiokol chemists set out to reduce the sulphur content of the polysulfide rubber binder used in their solid propellants. They experimented with several polymers and copolymers. Polybutadiene rubber polymers were found to be the best candidates.

Eventually, PBAA (a copolymer of polybutadiene-acrylic acid) was developed and used in the first stage of the Minuteman I ICBM (first introduced in 1962). PBAA was an improvement on the original Thiokol rubber binder, but lacked tear (tensile) strength.

However, the search for ever better rubber binders and casing binders continued. This led to a new synthetic rubber binder PBAN (butadiene terpolymer, acrylonitrile, and acrylic acid). Thiokol introduced PBAN in 1954. It had superior physical properties to PBAA, due to the addition of 10% acrylonitrile. Thiokol then developed CTPB (carboxyl-terminated polybutadiene). This was even better, but too expensive for widescale use. Thiokol binder evolution ended with the HTPB (hydroxyl-terminated polybutadiene) binder. This low-cost, low-viscosity propellant binder has become the new industry standard.

Thus polyurethane binders took over with the first long range solid-fueled ballistic missiles in 1960 with the Polaris A-1 SLBM, and then the Minuteman I ICBM in 1962. It was followed by Polybutadiene-acrylic acid (PBAA), then Carboxyl-terminated polybutadiene (CTPB), followed by Hydroxyl-Terminated Polybutadiene (HTPB) in 1976 with the Trident I SLBM. Finally there were cross-linked composite propellants, as well as cross-linked double-base propellants.

Ammonium perchlorate processing was as follows.

First the AP is dried by heating to 235 °F for at least twenty-four hours. If the AP is not at the required mean particle size (10% 80µ + 90% 5µ), it was sieved, and/or ground in a hammer-mill.

Then the fuel pre-polymer (monomer; binder) is thoroughly mixed with it curing agent into a thick paste, and then place in a propellant mixer whose walls are heated to a constant 135 °F. The weighed AP is then slowly added while mixing under remote control. After 2-3 minutes, the AP is thoroughly wetted, and vacuum is drawn, and mixing continued for another thirty minutes.

The propellant is cast, preferably under vacuum. For small size AP particulate, sometimes hand-packing is necessary to avoid voids in the propellant. Following casting, the propellant is cured at 176 °F for forty-eight hours. Catalysts such as copper chromite or copper (II) oxide promote the high temperature burning of AP. The addition of aluminum produces a high specific impulse. [Steinz (1969)]

The main function of a solid rocket nozzle is to channel and control the expansion of hot gases coming from the missile solid fuel chamber, thus creating the motor thrust. The most usual device for thrust vector control ("steering" the missile) is a flexible nozzle bearing allowing thrust vector angles from 3 to 15 degrees. The movable nozzle is hinged by a flexible bearing, a ball and socket arrangement, or a hydraulic bearing joint. The nozzle may have an extendible exit cone to increase the hot gas expansion ratio, and allows a 6 - 8% increase in thrust performance. [AD-A199 356 (1988)]

The nozzle is at the base of the missile casing, which is made of composite materials, originally glass fibers. Composites are "laminates" of a unidirectional ply of continuous fibers lying side-by-side, parallel, and essentially one fiber thick. The fiber is bound together by a resin matrix, which infuses the fibers.

With the Trident SLBM, the polymer resin is Bisphenol-A/Epoxy. The fiber was made of glass fibers, which then evolved to Kevlar-49 fibers, and then evolved to a coating of pyrolytic graphite (made by Hercules).

Each layer of fibers is very strong and stiff in the direction of the fiber, but comparatively quite weak in other directions. Thus there are several layers of glass fibers in different directions compared to the base layer, that provide full strength of the casing in all directions of interest. The laminate of layers of glass fibers in an resin matrix has stacked layers of "lamina" to achieve the desired strength or stiffness.

The solid propellant is in the form of "grains" are bonded by a liner to the solid propellant casing (case-bonded grains). The grains are bonded to the motor case during casting, and curing steps of the propellant grain manufacturing process. The outer surface of the grain is bonded to a casing liner (and a thermal insulation layer), whose primary purpose is to provide an adhesive bond between the propellant grain and the EDPM rubber internal casing insulation, which provides thermal protection to the case from combustion products, and also structurally supports the propellant grain within the motor case. The casing insulation is a filled rubber material, with typical fillers silica and Kevlar fibers.

The primary materials in a solid propellant grain (propellant, liner, and insulation) are basically nearly incompressible, rubber-like materials, with a bulk moduli of compressibility of 200 kpsi or greater in a virgin (undamaged) state. However, they respond generally as visco-elastic materials. [AD-A199 356 (1988)]

United States Patents chronicle the evolution of solid fuel propellants over the last twenty years:

(NC = nitrocellulose; NG = nitroglycerin; AP = ammonium perchlorate; Al = aluminum powder)

1959 (declassified 1973; Phillips Petroleum): 90 weight % AP solid propellant with 10% rubber binder. Natural rubber, synthetic rubber, epoxy resins, polybutadiene, and many others. With burning rate catalyst di- (2-(2-pyridyl)pyridine copper (II) dichromate. (U.S.P. #3,753,348) (U.S.P. #3,779,824) (U.S.P. #3,779,825).

1959 (declassified 1964; Klager, Aerojet-General): First polyurethane solid propellant patent; polymerization catalyzed with FeAA.

1959 (declassified 1966; Klager, Aerojet-General): Polyurethane binder resin of low temperature cure.

1960 (declassified 1975; Army): Composite Modified Double-Base (CMDB), a Triple-Base propellant; 50% HMX, 22% NG, 18% NC, plus the binder. HMX the oxidizer. Cross-linked for stability. Ingredients mixed under vacuum, then vacuum cast, and cured at around 150 °F until solidifies. Impulse of 14.0 lbs. per cubic inch when burned at a pressure of 1 kpsi (U.S.P. #3,878,003).

1960 (declassified 1979; Aerojet-General): manufacture of a bipropellant grain. (Used in the MM I ICBM at least.) (U.S.P. #4,137,286).

1960 (declassified 1973; Arco [Atlantic Research Co.]): Up to 15% by weight of aluminum powder of 3 micron particle size and AP is added to Double-Base (NC + NG) solid propellant (U.S.P. #3,779,826).

1961 (declassified 1967; Aerojet-General): Solid propellant (U.S.P. #3,296,043).

1961 (declassified 1974): Purification of HMX (U.S.P. #3,853,847).

1961 (declassified 1974; Rockwell): Use of solid fuel reinforcing aluminum or magnesium metal filament lattice to increase structural stability and tensile strength, and reduce the amount of polymeric binder required in the solid fuel (U.S.P. #3,811,358)  (15 patents reference this one in April, 2000).

1961 (declassified 1976; BASF):  Polyurethane binder for double-base propellants: new  propellant class:  cross-linked double-base composite (U.S.P. #3,956,890)  (17 patents reference this patent in July 2000).

1961 (declassified 1976):  Rubberized missile casing liner material (U.S.P. #3,965,676).

1961 (declassified 1970):  First Hydroxyl-Terminated Polyurethane composite propellant  (U.S.P. #3,532,567).

1962 (declassified 1975; Army):  Cast polyurethane rubber/AP propellant.  Polyurethane is the binder as well as the fuel for the AP oxidizer (U.S.P. #3,870,578)  (19 patents reference this one [in 1999]).

1963 (declassified 1971; Army) Up to 5% elongated ferromagnetic iron particles in propellant oriented with a magnet to accelerate burning rate, and thus increase thrust (U.S.P. #3,617,586).

1964 (declassified 1975; Navy):  Cast double-base propellant (60% NG + 15% NC) (U.S.P. #3,907,619).

1964 (declassified 1977; Navy):  Cross-linked Triple-Base Propellant;  26% NG, 9% NC, 40% HMX, 18% Al powder.

1964 (declassified 1974; Arco):  PEG (polyethylene glycol) as plasticizer for double-base  propellants.

1964 (declassified 1978; Hercules):  a rocket casing filled with radial metal filaments from the center of the casing before filling with propellant provides increased structural strength, as well as burn rate (Stage 2 of Polaris A3 made by Hercules in 1964)  (U.S.P. #4,085,173).

1964 (declassified 1976; Arco):  Aluminum staples dispersed evenly in double-base or composite fuel especially.  Staples are 0.375" long, up to 0.005" wide, and 0.001" thick.  21% aluminum (U.S.P. #3,933,543) (17 patents reference this one in 1999).

1964 (declassified 1978; Thiokol):  First patent for a CTPB (Carboxyl-Terminated Polybutadiene) composite, binder, with aluminum powder, and less than 3 micron (μm) size ammonium perchlorate (U.S.P. #4,070,212).

1964 (declassified 1978; Thiokol):  "High Performance Fast Burning Solid Propellant" (U.S.P. #4,070,212).

1965 (declassified 1977; Du Pont):  "Derivatives of Carboxy-Terminated Polybutadienes"; difluoroamino derivatives for higher specific impulse (U.S.P. #4,042,619).

1966 (declassified 1978; Aerojet-General); "Staple Orienting Method and Apparatus"; aluminum staples act as heat conductors, increasing the propellant burning speed; the stapes (foil, wire, etc.) is forced with the uncured propellant through a screen, to align the staples perpendicular to the surface of the burning propellant. (U.S.P. #4,103,584).

1966 (declassified 1974; Navy):  Castable polyurethane composite propellant; 65% AP + 15% Al powder + 20% polyurethane (U.S.P. #3,791,892).

1966 (declassified 1979; Navy):  Polyurethane binder solid propellant; 46% AP + 23.4% NG + 21.5% beryllium hydride + 9.1% polyurethane binder.

1967 (declassified 1975):  Double-base propellant burning pressure of 1 - 2 kpsi; burning rate of 0.25" per second (U.S.P. #3,868,282).

1968 (declassified 1974; Hercules): Cross-linked diisocyanate double-base propellant (U.S.P. #3,798,090).

1968 (declassified 1980; Hercules): "Composite Propellant with Surface Having Improved Strain Relief"; Advanced Sparrow Missile (U.S.P. #4,241,661).

1968 (declassified 1977; Hercules)(in 2000, 10 patents ref it): "Composite Modified Double-Base Propellant with Filler Bonding Agent"; (U.S.P. #4,038,115) Improved binding agent increases propellant's resistance to cracking and mechanical stress.

1969 (declassified 1981; Hercules): "Cross-linked Smokeless Propellants"; double-base propellant with good mechanical properties (e.g. no cracking) and Specific Impulse of 246 lbf.-sec/lbm consisting of 48.5% NG + 7.4% NC + 29.9% RDX (U.S.P. #4,298,411).

1970 (declassified 1973; Army): 16% double-base, 40% HMX, 30% AP, and 15% Al; cross-linked nitrocellulose propellant with isocyanate to prevent void formation.

1970 (Army): "Double-Base Propellants with Combustion Modifier"; catalyst effective in lowering the temperature dependency of burning rate with pressure; a 90% NC – 8% NG double-base powder mixed with HMX, and catalyzed with lead stannate TDDI complex (U.S.P. #3,951,704, declassified 1976).

1970 (declassified 1974; Aerojet-General): HTPB (Hydroxyl-Terminated Polybutadiene). Propellant with 86% (AP + Al powder) (in 2000, 11 patents reference it) (U.S.P. #3,801,385).

1970 (declassified 1975; Army): CTPB (Carboxyl-Terminated Polybutadiene) propellant with 80% (AP + Al powder) (U.S.P. #3,914,140).

1970 (declassified 1978; Aerojet-General): "Coated Ammonium Perchlorate" (U.S.P. #4,115,166).

1971 (declassified 1976; Thiokol): Coated, ultra-fine ammonium perchlorate, liquid-processed with a coating agent such as CTPB for a 0.8 micron (μm) size (U.S.P. #3,954,526).

1971 (declassified 1977; Thiokol): Method of making 1 micron (μm) ammonium perchlorate particles (U.S.P. #4,023,934).

1971 (declassified 1982; Navy): "Solid Propellant Compositions", 60 - 85% AP + 2 - 26% Al + S + 2% burning rate accelerator Cu chromite + polymeric binder + a cure catalyst Fe acetylacetonate; stable to thermal cycling, good storage stability, rapidly cured, good ballistic properties, and relatively safe (U.S.P. #4,332,632).

1971 (declassified 1987; classified 16 years, Army): Method of making ultra-high surface area ammonium perchlorate (possibly Spartan/Sprint propellant) (U.S.P. #4,698,106).

1972 (declassified 1980; Navy) "Stress Reducing Liner and Method of Fabrication"; coating of a rocket casing with curable polybutadiene, and heating and cure spinning it to form an even layer, coating the layer with perforated silicone rubber, coating, and spinning with a third layer of uncured polybutadiene. A liner between the case and propellant is well known to prevent stress fracturing, cracking, and attach the propellant with the casing (U.S.P. #4,185,557).

1973 (declassified 1984; Hercules): Urethane stabilizer for CMDB (Composite Modified Double-Base) propellant; 41% HMX + 19% NG + 8% NC + 20% 30 micron (μm) Al powder (U.S.P #4,478,656).

1973 (declassified 1984; Hercules): CMDB Propellants stabilized with urethane compounds; 8.2% NC + 19.1% NG + 40.9% HMX + 20% 30μ Al powder + 8% AP (U.S.P. #4,478,656).

1973 (declassified 1978; Thiokol): "Method for Mounting Solid Propellant in a Rocket Motor".

1974 (declassified 1988): "Extrusion Method for Obtaining High Strength Composite Propellants"; Navy; Cross-linked composite propellants. Replaces casting of these propellants with extrusion (U.S.P. #4,776,993).

1975 (declassified 1988; Goodyear): Polyurethane propellant with superior physical properties (U.S.P. #4,753,751).

1977 (declassified 1995; Hercules): Substituted para-nitroaniline stabilizers are disclosed for isocyanate cross-linked CMDB propellants (U.S.P. #5,387,295).

1976 (declassified 1990): Composite double-base propellant mixed with small aluminum staples: porous AP + Al staples + NG + NC + Al powder. Ultra-high burning rate. (U.S.P. #4,944,816).

1979 (declassified 1992): "Composite Solid Propellant with a Pulverent Metal/Oxidizer Agglomerate Base" (U.S.P. #5,139,587).

1980 (declassified 1990; Navy): Composite propellant: 80% oxidizer (HMX or AP) + 10% metal fuel (Al or Mg powder) + 10% polyurethane polymer binder. (U.S.P. #4,944,815).

1983 (declassified 2000): Solid Propellant.

1983 (declassified 1987; Hercules): XLDB (Cross-linked Double-Base) propellants with HMX or RDX oxidizer of small (15 μm) particle size creates dramatically enhanced burn rates (U.S.P. #4,689,097).

1993 (declassified 2015): Solid propellant bonding agents (U.S.P. #9,181,140).


## Solid-Fueled Inter-Continental Ballistic Missiles (ICBM)

| ICBM | First Deployed | Range (km) | Warhead | Yield |
|---|---|---|---|---|
| Minuteman I (first solid-fueled ICBM) | 1962 | 10,560 | W-59 | 870 kt |
| MM II | 1966 | 12,700-14,400 | W-56 | 1.27 MT |
| MM III (first MIRVed ICBM) | 1970 | 14,400 | 3 x W-62 | 3 x 170 kt |
| MM III Upgrade | 1979 | 14,400 | 3 x W-78 | 3 x 335 kt |
| MX/Peacekeeper | 1986 | 11,000 | 10-12 x W-87 | 10-12 x 335 kt |

[Mackenzie p. 167, 205, 241 (1993)] [Stumpf, p. 185 (2000)] ["Minuteman Weapon System" (2001)] [OTA, Appendix B (1981)] [Goldberg, part 2, p. 572 (1981)]


The 1,000 silos of the Minuteman ICBM (for the MM II & III) formed the backbone of the U.S. strategic missile force, starting the eclipse of the SAC B-52 bomber force in 1962. They had speed (time to target of 30 minutes), accuracy (superior CEP of bombers and SLBM's), and numbers (450 single warhead MM II, and the 550 x 3 MIRV warhead MM III). The Minuteman was deployed in 20 strategic missile squadrons, each possessing 50 missiles. Three to four squadrons were assigned to each of six Wings – Wings I, II, III, IV, V, and VI.

Minuteman I (SM-80, renamed LGM-30A & B; WS-133A ground system) (October 1962)

Warhead: MM I-A (1962): 870 kt W-59 (U238 secondary); MM 1-B (1965): 1.27 MT W-56 (Oralloy secondary); ICBM throw-weight: 1,000 lbs.

RV: AVCO Mk 5 then the Mk 11; 18" dia. x 18" high. The development of the much more flexible, reliable, and safer solid-fueled MM I allowed the early retirement of the first generation, liquid-fueled Atlas and Titan I ICBM's.

Stage 1:

Casing: 0.15" thick Ladish D6AC steel (ultra high-strength, low-alloy, martensitic steel); 220 kpsi tensile strength; double vacuum melted; Tempered for 2 hrs. at 900 °F. [AD264 660 (1961)] The casing used high strength English Sheffield strip steel (essentially 1095 razor blade stock, or clock spring steel) [Dorman (1995)] [AD701 545 (1969)]
Thrust: 167,000 lbs.
Propellant: 20.8 tons; 67.5% AP + 10% Al powder + 22.5% PBAA (polybutadiene-acrylic acid binder) with a burning rate of ~0.3"/sec at 1,000 psia [AD-A952 021 (1961)]; 6 pointed, star-shaped (6 arms) cavity all the way through propellant
TVC: (Thrust Vector Control) four nozzles pivoted in pairs via gimbals, plus or minus 8° rotation, with phenolic Refrasil as insulation/heat protection. The nozzle blast tube was made of ADHG graphite, with a tungsten nozzle throat (possibly with 15% molybdenum alloyed in, which was best for tungsten nozzles). The unalloy tungsten (m.p. 6,200 °F) castings were machined (pressed, sintered, and then forged) from as-cast vacuum-arc-melted ingots, or by centrifugal castings to obtain finer grain size. [AD268 311 (1961)].
Contractor: Thiokol Chemical Co., Brigham City, Utah (Aerojet-General secondary contractor).

Stage 2:

Casing: Initially used steel casing, then replaced by a high-strength, heat-treated, solution-treated-and-aged, beta-titanium alloy casing (Ti-6Al-4V); 160 - 180 kpsi tensile strength; the use of Ti-6Al-4V was limited to the second stage of the MM II because of cost [AD818 416 (1967)]; the ignition temperature of titanium is 1,300 °C. [AD-A102 516 (1981)]
Thrust: 49,200 lbs.
Propellant: 82% AP (+ polyurethane binder) to 88% AP (+ polybutadiene binder)
TVC: Swiveling TVC; 4 swiveling nozzles, plus or minus 6° gimbal; Nozzle Blast Tubes: Ti coated with MX2630A carbon-phenolic, with ATJ graphite nozzle inlet, tungsten in ATJ ring, with molybdenum metal sleeve for nozzle throat, and nozzle exit cone of MX2625 graphite-phenolic.
Contractor: Aerojet-General, Sacramento, California (Thiokol secondary contractor)

Stage 3:

Casing: Fiberglass-epoxy (S-994 Glass/ERLA 2256 epoxy)
Thrust: 18,900 lbs.
Powered flight: 15.4g's
TVC: 4 swiveling nozzles, with plus or minus 4° maximum gimbal angle; Nozzle inlet ATJ graphite, with Nozzzle throat tungsten metal in a graphite ring, and nozzle rings of graphite.
Propellant: Diisocyanate cross-linked modified double-base (35% NG + 11% NC) + 22% AP + 20% Al powder + 11% triacetin + 1% nitrodiphenylamine. [U.S.P. #3,894,894 (1975)]
Contractor: Hercules Powder Co., Magna, Utah.

Weight of the three Minuteman I motors almost double that of the entire two stages of the Polaris A-2.

Stage 3 Warhead Nose Fairing:  133" long x approx. 19" diameter

The MM I (LGM-30A) ICBM:

Height:  53.7' total height
        1st stage 65" diameter x 23.6' long;
        2nd stage 44" dia. x 10.5' long;
        3rd stage 37" dia. x 7' long.
Weight:  65,000 lbs.
Ceiling:  700 miles
Range:  LGM-30A:  4,900 nm
        LGM-30B:  5,500 nm   [Goldberg, part 2, p. 572 (1981)]
Speed:   Mach 19.7
Accuracy:  1.0 nm CEP
Number:  By July 1964, there were 600 MM I's in service.  [Neufeld (1990)]; 800 by the end of June 1965. ["Minuteman Weapon System" (2001)]  By September 1974, the last MM I was retired.
Overall Contractor:  Boeing

Bi-propellant grain; Titanium engine chamber; Durable nozzle throat; graphite throat insert to solve cracking problem from thermal shock; shrink-fit Molybdenum-ring, and later, a forged Tungsten insert.

Alclo (Al powder + potassium perchlorate) rocket igniter used in all U.S. Minuteman missiles (non-gassing heat source).

The NS10 MGS (Missile Guidance System) used a magnetic core memory, the first type of practical computer memory.  It consisted of thousands of tiny ferrite doughnuts threaded on two wires each in a three-dimensional matrix of fine wires and ferrite doughnuts reminiscent of a spider web.  The cores were magnetized by the direction of a flow of current either clockwise or counter-clockwise to represent a bit, either "0" or "1".

The magnetic core memory is unfortunately DRO (Destructive Read Out), which means its contents are erased by being read, so must be re-initialized after its contents are read out.  It is also expensive and tedious to thread the tiny ferrite cores on the wire matrix, which must be done by hand.  But it worked, until technology came up with a better and faster form of RAM (Random Access Memory).  [Foster, p. 25-29 (1985)]

Magnetic core memory was improved immeasurably and ingeniously by gluing a two-dimensional array of ferrite cores on a thin plastic card, many of which were stacked together for greater capacity.  [U.S.P. #2,784,391 (1957), filed 1953]

The Minuteman program, from its very inception, was oriented towards mass production of a simple, quick, efficient, and highly survivable ICBM, that was inexpensive to maintain and operate.  Colonel Edward Hall, Chief of Propulsion Development for the Western Development Division was selected by General Bernard A. Schriever, Commander of the W.D.D. to head the program to develop the solid-fueled Minuteman program.

The MM I's were based in underground silos across the mid-west (listed from South to North):  Arkansas, Kansas, Missouri, Colorado, Nebraska, Wyoming, South Dakota, Montana, and North Dakota.  [Arkin, p. 48 - 49 (1985)]  They were placed in the central U.S. to avoid a surprise attack on the silos by Soviet submarine launched SLBMs.

[Stine (1991)] [Berman (1983)] [Braun (1975)] [Hansen (1988)] [Sheehan (2009)] [Baker, p. 187 (1978)] [Strategic-Air-Command.com (ca. 1995)] [National Park Service (ca. 1995)] [Hove (1979)] [AD355 925 (1963)]

Minuteman II (LGM-30F) (Oct. 1965)

1.27 MT, 564 lb. W-56 warhead ["U.S. Strategic Objectives" (1971)] [Hansen (1988)]

RV:  hemisphere-cylinder-flare (skirt) RV (cylinder 18.8" diameter; 64.1" RV total length); re-entry speed of 7 km/s [AD-A224 584, p. 4-6 (1990)]

1,100 lbs. throw-weight  [Buchonnet (1976)]   There are 8 possible targets stored in advance in the guidance computer.  [DNA 6147T (1983)]

The MM II used the same Stage 1, 2, and 3 missile casings as the Minuteman I.


Stage 1:

Casing:  Same as MMI, Ladish D6AC steel; 220 ksi; Thiokol
Propellant:  AP + Al powder + polybutadiene-acrylic acid (PBAA) fuel/binder with epoxy resin curing agent; 6-point star     hollow core
TVC:  four swiveling nozzles; same as MM I.


Stage 2:

Casing:  Titanium casing (same as MM I); Aerojet-General
Propellant:  78% AP + 10% Al powder + 12% CTPB (Carboxy-Terminated Polybutadiene) rubber binder [U.S.P. #3,948,698 (filed 1967; declassified 1976)]
TVC:  1 fixed submerged nozzle; liquid injection TVC; Nozzle inlet:  molded MX2630 carbon-phenolic; Nozzle throat:  W coated with ~ 80 mils fine grain, Grade ATJ pyrolytic graphite, and with Mo (molybdenum)  sleeve; Nozzle exit cone:  tape-wrapped FM5014 graphite-phenolic with silica-phenolic extension.


Stage 3:

Casing:  Glass fiber-epoxy composite (same as MM I); Hercules
Propellant:  Solid double-base
TVC:  Nozzle inlet:  graphite-phenolic; Nozzle throat:  W with pyrolytic graphite rings.


New second stage motor to increase range over MM I, with a secondary liquid injection system of TVC. Also four thrust termination assemblies, at the top end, which explosively blow off, and separate the third stage from its RV, and cause it to back away from the RV.  [AD-A363 899 (1985)]

Total Weight:  74,000 lbs.
Size:  1st stage:  65" dia. x 24' long;
          2nd stage:  52" dia. x 14' long;
          3rd stage:  37.9" dia. x 7' long
Speed:  Maximum speed in excess of 16,000 mph
Guidance:  NS17
**CEP:  0.43 nm** ["U.S. Strategic Objectives" (1971)]

Range: 7,000+ nm
Number: 450

In May 1969, there were 500 MM I and 500 MM II. [Neufeld (1990)]


Minuteman III (LGM-30G) (1970)

("L" designates silo-launched; "G" denotes "ground attack" and "M" means "guided missile")

First MIRV'ed ICBM; first conically-shaped RV's: three Mk 12 RV's with W-62 170 kt warheads [NSC (1971)]

W-62: warhead 7" dia. x 22"; Barnaby claims W-62/W-78 weight 100 kg; Greenwood (1988), p. 169 says (Mk 12 RV + W-62 warhead) weighs 400 lbs. (182 kg); Bunn (1984) says 200 kg.

RV: First used an Avco Mk 11C (32" dia. x 97" high; 9.4° half-angle; Avco RAD 60 silica-phenolic nose tip with an OTWR silica-phenolic RV heatshield; Dynasil quartz antenna window), then soon replaced by:

GE Mk 12 (20" dia. x 42" high; 0.5" dia. hemispherical RV nose tip; 8.3° half-angle; 700 - 800 lbs.; Teflon over inverted Dixie cup carbon-phenolic 5055A nose tip with carbon-phenolic 5055A RV heatshield; slip-cast forged silica antenna window) [Stine (1991)] [AD-A955 400, p. 16-40 (1972)]

Mk 12 MIRV aeroshell for the W-62: frustum of a cone (a conical segment truncated at the top of the cone), 20" dia. at the base x 38.5" high [SAND97-8012 (1998)], and 8" in diameter at the top. The aeroshell structure was 0.51" thick (0.41" thick outer layer of carbon fiber-reinforced phenolic, glued with a 0.040" thick Epon 934 layer to a 0.060" thick 7049-T7352 aluminum inner layer), making the cavity dimensions 22" high, and 7" in diameter at the top. The aeroshells that contain the W76, W78, W87, and W88 devices are of similar construction. [SAND94-0489 (1994)]

Without the bottom portion (of a possible 3" height), with a little trigonometry, this means the Mk 12 RV is at least 41" high (44" with the 3" base), with a diameter of at least 17.3", an apex half-angle of 12° (total apex angle 24°), and a carbon aeroshell weight (at 2 g/cm3) of 29.1 kg or 64 lbs. The nose cone may be ½" in diameter, with a half-angle of 10°, and made of 0.5 – 8% boron-doped pyrolytic graphite ("ductile" graphite). [AD824 697 (1967)] [AD392 561 (1968)]

The Mark 12 RV used an inertial fuzing system. An accelerometer measures the drag deceleration and integrates it over time. When a predetermined value of the integral is reached, at an altitude called the signal altitude, a timing device starts counting. After a predetermined time interval has elapsed, the warhead detonation mechanism is activated. [Strategic-Air-Command.com (ca. 1995)] [Hove (1979)] [AD-A197 073 (1988)] [AD353 247 (1964)]


Stage 1:

Casing: Ladish D6AC steel; 220 ksi; same as MM I & II.
Propellant: AP + 60 µm diameter or less, spherical Al powder + polybutadiene acrylic acid binder with epoxy resin curing agent; 6-point star hollow core; 60 second burn; propellant has a specific impulse of 250 seconds (as opposed to 450 seconds for liquid propellant of the Space Shuttle (H2 + LOX)); Thiokol M55.
TVC: Four movable gimbaled nozzles; nozzle: AISI 4130 steel with forged tungsten, pyrolytic graphite-insulated throat.
Thrust: 202,600 lbs. (91,170 kg) for 60 seconds
Contractor: Thiokol

Stage 2:

Propulsion:  Aerojet SR19
Casing:  Titanium case (Ti-6Al-4V; 160-180 ksi); same as MM I & II.  Uses CTPB binder in casing liner. Also uses HX 868 (2-ethyl aziridine amide with trimesic backbone structure) as a CTPB cross-linking agent. [AD-A387 318 (1995)]
TVC:  1 Fixed Submerged Nozzle; liquid hot gas injection TVC; same nozzle as MM II.
Thrust:  60,700 lbs.; 60 second burn
Contractor:  Aerojet-General


Stage 3:

Propulsion:  Aerojet-Thiokol SR73
Casing:  S-994 Glass/ERLA 2256 epoxy; same as MM I & II.
Propellant:  a CMDB (Composite Modified Double-Base) propellant consisting of 45% HMX (180 µ particle size), 25% NG, 7% NC, 18% 60 µm spherical Al powder, 4% diisocyanate cross-linker, 1% 2-nitrodiphenylamine stabilizer, and 0.005% dibutyl tin diacetate catalyst.  Uses CTPB propellant binder in missile casing liner. [AD-A387 318, p. III-104 (1995)] [U.S.P. #3,716,604, Hercules (1973)]  S-901 fiberglass made solely by Owen-Corning, used in both Polaris A-3 and the 3$^{rd}$ stage of MM III,  with resin impregnation by Ferro Corp, prior to chamber fabrication.
TVC:  Fixed nozzle with liquid-hot gas injection (similar to 2nd stage MM II nozzle) to increase range. Nozzle the same as in the MM II, except exit cone MX4926 carbon-phenolic with MXC-113 carbon felt tape extension.
Thrust:  34,876 lbs. for a 60 second burn

Contractor:  United Technologies Co., Chemical Systems Division
Managing Contractor:  Boeing

PBV (Post-Boost Vehicle/"Bus"):  "There are four primary subsystems which make up the Post-Boost Propulsion System:  gas pressurization, propellant tank with positive expulsion, controls, and engines. The total system supplies propellant ($N_2O_4$/MME [Nitrogen Tetroxide/Monomethyl Hydrazine]) on demand to the attitude control engines for control of post-boost vehicle attitude.  It also provides this on-demand propellant to the axial thrust engine(s) to permit changes in vehicle velocity to ensure proper vehicle re-entry position as well as over-all velocity control."  Propellant is stored in two equal-volume, conospheroidal tanks.

Fluidic Controls:  This effort was undertaken by the Bowles Engineering Corporation, Silver Spring, Maryland. The concept under development utilizes all-fluidic components for both the regulation of pressurization gas and the control of bipropellant engine flow.  The fluidic concept offers a number of advantages in terms of inherent reliability as well as long-term storage because there are no moving parts.  The design requirements of the fluidically-ccntrolled PBPS dictated a two-stage monopropellant $N_2H_4$ decomposition gas generator subsystem.  [AD392 715 (1968)]

6 pitch/yaw motors, and 4 roll motors, all liquid-fuelled, are used for MIRV bus directional control; a 23 lb. thrust pitch and yaw control engine; with 0.28 lbs. of fuel (monomethyl hydrazine), and 0.40 lbs. of oxidizer (nitrogen tetroxide), which gave 6 seconds of burn time, helium was used for pressurization and purging the system,  [AD-A286 599, p. 74 (1972)] [AD-A258 264 (1992)]  and used a 0.030" thick steel bellows oxidizer pump on the PBV tanks. The nitrogen tetroxide contained a reduced concentration (over the Titan II ICBM) of 0.11 – 0.51% NO (nitric oxide) to control (reduce) its corrosion of the thin (30 mil thick) steel bellows. The solution of NO in the nitrogen tetroxide is called MON (Mixed Oxides of Nitrogen). [AD-A036 741 (1977)]

Its total weight was 3,650 lbs.  Its dimension were 95" in diameter x 28.5" high. It total impulse was 684,000 lb.-sec. (632,000 axial and 52,000 Attitude Control System (ACS)) (its Specific Impulse was 300 sec. for axial and 285 sec. for ACS).  Its axial thrust was 600 lbs. and its ACS thrust was 75 lbs. Capable of axial, pitch, yaw, and roll maneuvers, in order of decreasing magnitude.  The mission time was 800 seconds.  [AD392 715 (1968)]


Grain Design (for the 3 solid fuel stages):  Cylindrically perforated
Size:  59.9' total length;
    1st stage:  24' high x 65" dia.; 200,000 lbs. thrust for 60 seconds; Thiokol
    2nd stage:  14' high x 52" dia.; 60,000 lbs. thrust; liquid-injection for TVC; Aerojet General
    3rd stage:  7.3' high x 52" dia.; 34,876 lbs. thrust; internal case pressure:  600 psi; Thiokol & Aerojet General
    MIRV Shroud:  12' high x 52" dia. (base)

A silver-zinc battery, which provides 28 volts dc, powers the Flight Control System, the NS20 MGS, and the Reentry System

Weight:  79,432 lbs. (32,158 kg)
Throw-Weight (nukes + RV's + PBV):  4,432 lbs. ["Strategic Forces Technical Assessment Review" (1983)]
Maximum Range:  8,083 m
Maximum Speed:  15,000+ mph (Mach 23 or 24) at 3rd stage burnout
Apogee (Ceiling):  700 miles (1,120 km)
**CEP:  0.25 nm** ["U.S. Strategic Objectives" (1971)]; 0.1 nm [Bennett (1980)]  Compared to the older MM2, the range of the MM3 is about the same, as well as the CEP being roughly the same. [JCAE (1973)]
Number:  550

[USAF Fact Sheet 96-09 (1996)]  [Strategic-Air-Command.com (ca. 1995)]  [OTA, App. B, p. 328-9 (1981)] [Hove (1979)] [Goldberg, part 2, p. 875 (1981)] [AD-A511 972 (2009)]



Minuteman III (Upgrade) (1979)

Upgrade of MM III from 1979 - 1983.

The Mk 12 RV's were upgraded to the improved Mk 12A RV's (with a carbon-carbon nose tip, and a 0.5" thick, carbon fiber reinforced-phenolic 5055A heatshield over a thin aluminum aeroshell substrate) with new warheads with double the yield, the 335 kt W-78 warhead, replacing the old W-62.  "The Minuteman III Intercontinental ballistic missile (ICBM)…can be retargeted remotely from the Launch Control Center." [DNA 6147T, p.1 (1982)]

Total Weight:  78,000 lbs.
Throw-weight:  2,400 lbs. [Nitze (1979)]
Propellant:  Third Stage: AP + 20% Al solid propellant + HTPB binder [AD786 472 (1974)]
Guidance:  NS20
CEP:  0.12 nm
Underground silo hardness:  2,000 psi



MX/Peacekeeper (LGM-118A) (December 1986 – 2005)

10 – 12 W87 warheads, in one of two MIRV configurations; yield alleged to be 335 kt, upgradeable (but never was) to 475 kt.  The Oralloy required to up the yield was won over by, and diverted to, the competing Trident II W88 SLBM's warhead.

10 or 11 Avco Mk 21 ABRV's (Advanced Ballistic Reentry Vehicle), or 12 W78 warheads in smaller diameter Mk 12A RV's (9 around the circumference, with three centrally positioned).  The START II Arms Control Treaty later limited the number of RV's on the MX to 10.  The MX was designed as a large, long range, heavy throw-weight ICBM, just to (stupidly) match equivalent heavy throw-weight Soviet ICBM's.

REB (Reentry Body) (Mk 21 ABRV + W87) weight:  about 465 lbs. (210 kg); W87 alone guesstimate 150 kg.

Throw-weight:  7,940 lbs. (3,600 kg), also listed as 8,710 lbs.  Included 11 REB's + PBV weight of 3,000 lbs. [Burton (2013)] (which included 1,400 lbs. of PBV liquid propellant).  [AD-B060 927 (1981)]

Dimensions:  92" dia. x 71' high; diameter constant at 92" all the way up the three-stage missile to the MIRV platform, whose PBV/bus was considered the fourth stage of the ICBM.
Total Weight:  192,300 lbs. (87,400 kg) (almost three times the weight of the MM III).
Missile Casing:  Kevlar-49/epoxy filament wound casing for the first 3 solid propellant stages.
Velocity:  6.7 km/s
Range:  6,000 nm (somewhat less than the MM3)
Boost Phase:  Lasted 180 seconds, ending at a height of 200 km
End of Busing Phase (i.e., launch of all MIRV's):  After 650 seconds, at a height of 1,100 km, just before the apogee.
Apogee:  1,200 km

Stage 1:

Propellant:  49,000 kg; 599,000 lbs. of thrust; Thiokol Corp.
Nozzle Throat diameter:  23"
Nozzle expansion ratio:  1:12

Stage 2:

Propellant:  27,000 kg; HTPB binder with HX 868 for HTPB liner adhesion.  [AD-A387 318 (1995)]; 275,000 lbs. of thrust; Aerojet-General
Chamber pressure:  Stage 1 & 2:  1,000 psi; Stage 3:  700 psi
Contractor:  Aerojet-General

Stage 3:

Propellant:  7,700 kg; 65,000 lbs. of thrust; Hercules Corp.

PBV:  The Rocketdyne Corp. restartable MIRV "bus" was considered the "4th stage" of the Peacekeeper ICBM.  Its propellant was monomethyl hydrazine/nitrogen tetroxide (533.7 lbs. $N_2H_3CH_3$ and 866.3 lbs. $N_2O_4$) storable liquid  hypergolic bi-propellant delivered in equal volumetric flow-rates by 300 psi pressure from a gaseous mixture of helium and steam (from catalytic combustion of oxygen, and hydrogen gases.  MMH is a clear, nitrogen/hydrogen compound with a "fishy" smell. Nitrogen tetroxide is a reddish fluid and has a pungent, sweetish smell.  The propellants were stored in Kevlar 49/epoxy composite-wrapped (0.44" thick composite) 1100-0 aluminum alloy tanks which weighed 59 lbs. in total.  Each of the propellant tanks had an unpressurized propellant capacity of 10.57 cubic feet.  The pressurizing gas was in a spherical tank of volume 2,910 cubic inches at a pressure of 4,000 psia. [AD-B060 927 (1981)]

CEP: 90 m (as a result, many considered it a counter-force, first strike weapon)

Guidance: 115 lb. gimballess AIRS (Advanced Inertial Reference Sphere) gyroscope [Aldridge (1983)]; significant advance in guidance technology over Minuteman III to give the MX much greater accuracy. [OTA, App. B, p. 328-329 (1981)] The inertial measurement unit (IMU) is cooled by gaseous R-12 refrigerant, which is supplied from a reservoir within the flight coolant assembly. R-12, or dichlorodifluoromethane, is an organic compound of the CFC (chlorofluorocarbon) family. R-12 is commonly used as a refrigerant because of its nontoxic and nonflammable properties. It is readily converted from a gas to a liquid and vice-versa. ["Environmental Impact Statement Peacekeeper Missile System Deactivation and Dismantlement" (2000)]

Number: Only 50 were built.

Retirement: The last MX missile was retired in September 2005, in accordance with the START Treaty.

During its design, development, and build phase, it was originally called the MX ("Missile, Experimental"). It was then renamed by Ronald Reagan, with unintended irony, the "Peacekeeper." It was a heavy throw-weight ICBM designed for ten 475 kt MIRV's, that was meant to match the Soviet heavy throw-weight SS-18 ICBM, which was first fielded in 1974. It was considered a four stage ICBM: 3 solid-fueled stages + 1 liquid-fueled PBV.

To reduce aerodynamic drag (and thus increase range), the missile did not have the usual fins for stability, but instead used a sensitively-adjusted, single gimbaled nozzle. It was "cold"-launched from a reinforced steel canister as the silo. The missile was protected by Teflon-coated polyurethane pads (9 rows thick) that fell away on "launch" from the canister. It was ejected from the canister by 130 gallons (500 liters) of water turned to steam by a Launch Ejection Gas Generator (LEGG), a small rocket motor. [Burton (2013)]

It was originally to be a railroad-mobile ICBM in Utah and neighboring states for survivability, but tremendous local opposition killed that option, forcing the DoD to leave it stationed in fixed location silos. Then it had to fight over Oralloy allocation with the Trident II SLBM W-88 warhead. The W-88 warhead won, getting a yield of 800 kt, while the ultra-accurate MX's W-87 warhead was reduced from 475 kt to 335 kt. The MX also cost an exorbitant $68 million versus $25 million for the Trident II in 1989 dollars.

[Parker (2003)] [Cochran, p. 121-127 (1984)] [SAND95-8004, p. 81 (1995)] [SAND97-8017 (1997)] [Aerojet-General, p. IV-32 (1995)] [Francis, p. 180-181 (1995)] [Drake (1967)] [Burton (2013)]

## Submarine-Launched Ballistic Missiles (SLBM's)

| SLBM | First Deployed | Range (nm) | Warhead | Yield |
|------|---------------|-----------|---------|-------|
| Polaris A-1 | 1960 | 1200 | W-47Y1 | 400 kt |
| Polaris A-2 | 1962 | 1500 | W-47Y2 | 1.2 MT |
| Polaris A-3 | 1964 | 2500 | 3 W-58 MRV's * | 3 x 200 kt |
| Poseidon | 1971 | 2500 | 10-14 W-68 MIRV's ** | 40 kt each |
| Trident I (C4) | 1979 | 4000 | 8 W-76 MIRV's | 100 kt each |
| Trident II (D5) | 1989 | 6500 | 8 W-88 MIRV's | 800 kt each |

* MRV's (Multiple Reentry Vehicles) were for a single target, unlike the later (1970) MIRV design
** 14 W-68 MIRV's maximum

For safety reasons, the Navy insisted that all FBM's (Fleet Ballistic Missiles/SLBM's) be solid-fueled.

[MacKenzie, p. 241 (1990)] [Boyne (1998)] [Cochran, p. 142 (1984)] [Hearings I, p. 5] [Braun (1975)] [SAB2001796700, p. 23 (2001)]

Polaris A-1 (UGM-27A) (November 1960)

Classified technically as a IRBM (1,500 – 3,000 miles), for its limited range, it was the first of the 6 increasingly improved generations (ending with the Trident II) of the Fleet Ballistic Missile (FBM). First solid-fueled SLBM or ICBM, beating the Minuteman I by 2 years. Its limited range restricted it to the Mediterranean and the Norwegian Sea to be close enough to strike the eastern part of the USSR. [Mackenzie, p. 143 (1990)]

The W-47Y1 (400 kt) which weighed 600 lbs., had an integral beryllium, hemispherical nose – cylindrical body - flange/conical skirt, heat sink RV, weighing 235 lbs.; missile warhead compartment 18.26" dia. [AD388 603 (1957)] [Spinardi (1994)]

16 missiles for the Polaris submarine, of which there were 41 (for 656 missiles total).

Two-stage SLBM; First stage: Aerojet-General; Second stage:  also Aerojet-General
Dimensions:  54" dia. x 28.5 ft. long
Total Weight:  28,800 lbs. (First Stage: ~18,778 lbs.) (Second Stage:  ~9,469 lbs.)
Range:  1,000 nm
Altitude:  350 nm
Speed:  Mach 15

First Stage:

Propellant Weight:  16,300 lbs.
Propellant Specific Impulse:  240 s
Chamber Pressure:  1,000 psia
Total Impulse:  $3.8 \times 10^6$ lb.-sec
Average Thrust:  63,000 lbs.
Burn Time:  60 s

Second Stage:

Propellant Weight:  7,600 lbs.
Propellant Specific Impulse:  240 s
Chamber Pressure:  400 psia
Total Impulse: $2 \times 10^6$ lb.-sec
Average Thrust:  31,000 lbs.
Burn Time:  65 s  [AD388 603 (1957)]
CEP:  1.0 nm [Power (1964)]
Guidance Weight:  205 lbs.
Navigation:  Draper Ship Inertial Navigation System (SINS)

Missile Casing:  low alloy (AISI 4130) steel, rolled and welded; developed especially for Polaris
Fuel:  67.5% AP + 10% Al powder + 22.5% polyurethane binder (made using nitromonomers).  The aluminum powder was an Atlantic Research Corp./Aerojet-General innovation which significantly

increased the propellant's specific impulse. Same propellant for both stages, except ratio of AP:Al varied. Six-pointed star configuration. Propellant density was 0.061 lb. per cubic inch.

TVC (Thrust Vector Control): Four nozzles, each consisting of a steel shell, a single-piece molybdenum throat inserts backed by a graphite heat sink, and an exit-cone liner of a sandwich of molded silica-phenolic (nominal 32% phenolic resin) between steel and the molybdenum layers.

Jetevator: a solid ring with a spherical inside surface, which when turned into the exhaust stream, deflected the flow for steering. Made of a molybdenum alloy that was brittle below 400 °F, causing problems that were eventually resolved. Silver-infiltrated tungsten billets used in the fabrication of the nozzles. The billets were in the form of thick-walled hollow cylinders 6 - 9" high, 10" O.D., and 4" I.D., 15.2 - 16.3 g/cm3 (~80% density) and manufactured by Aerojet-General [AD295 888 (1962)] [ADA-080 547 (1979)]

Thrust Termination: Opening of six vent ports into the front of the second stage, sealed with plugs that could be blown out pyrotechnically.

Prime Contractor: Lockheed Corp. Other Contractors: GE, MIT, and Hughes (guidance)

Launch Method: The first launch of a missile from a submerged submarine. The bare missile was ejected with compressed air, directly into the sea water from the submerged submarine. The launch tubes were heavy-walled steel, machined smooth, to accomplish this, with the missile held snugly with three rings of flexible fall-away plastic pads.

Polaris stage separation means: U.S.P. #3,903,803 (filed 1960, declassified 1975). The most frequent problem encountered during testing was the loss of the carbon throat and exit cone liners from the nozzles. "Eventually molybdenum throat inserts backed by a graphite heat sink and molybdenum liners were used [for both stages] and worked satisfactorily." [Hove VI-4 &5 (1979)]

Polaris A-2 (UGM-27B) (1962)

Main changes: double the warhead yield; steel casing replaced with fiberglass-epoxy in second stage. In the second stage, the casing and aft opening was filament wound fiberglass, while the top end closure was a pressurized fiberglass filament laminate wound dome structure. The contour used was such that the longitudinal and circumferential glass filaments were stretched equally. The length of the cylindrical section of the second stage was 41.51", with a total length of 75.72", and an aft opening diameter of 7". [Siuta (1962)]

W-47Y2 (1.2 MT) Oralloy secondary [Francis, p. 138 (1995)] Compared to the Polaris A-1, the A-2 had increased reliability, but accuracy declined slightly. Three Polaris submarines normally patrolled the Mediterranean. [SAB2001796700]

Two-stage SLBM
Dimensions: 54" dia. x 31 ft. high (2.5 ft. higher than Polaris A-1)
Weight: 32,500 lbs.
Range: 1,500 nm

First Stage:

Same as Polaris A-1, but lengthened by 30"
Propellant: 67.5% AP + 10% Al powder + 22.5% polyurethane binder. Twelve-pointed star configuration.

Second stage:

Propellant: new cast-in-case, double-base (NC + NG) propellant, modified by addition of AP and Al powder

Missile Casing: For second stage, the casing was changed (from the Polaris A-1) from steel to S-901 fiberglass/epoxy (filament wound) (with over 3x the strength to weight ratio of steel), with significant weight savings. [AD-A080 547 (1979)]

TVC: 2nd stage jetevators were replaced by rotatable nozzles, which were more efficient and which gave TVC with a smaller loss in axial thrust. Gimbaled swiveling nozzle, with spherical ball-joint. [U.S.P. #3,912,172, filed in 1971, declassifed in 1975] [U.S.P. #4,350,297, filed in 1962, declassified in 1982] The A2 substituted pyrolytic graphite for molybdenum on the second stage.

Accuracy: 1.5 nm CEP [Power (1964)]

Polaris A-3 (UGM-27C) (1964)

Main changes: greatly increased range with same size missile required, so change in propellant; first stage also changed from heavy steel to filament-wound fiberglass-epoxy, with a first stage weight of the chamber of 1,455 lbs. [AD-A307 510 (1963)] First stage fired with increased temperature and pressure, so changed first stage nozzle with silver-infiltrated tungsten throat nozzle, the silver boiling off and cooling the nozzle.

3 x 200 kt W-58 MRV (equivalent to single 1 MT warhead) (First and only MRV ever fielded). Mk 2 RV + W-58 warhead weighed 309 lbs. [Hansen (1988)] The RV was of the hemisphere-cylinder-flare ablative type with a height of 54" and flare diameter of 23.5", and was integral to the warhead. The entire warhead was enclosed in a magnesium casing and an aluminum cover. The warhead RV was an ablative heat shield of nylon-phenolic laminate (a charring ablator) bonded to the warhead's magnesium case. [RS 3434/39 (1968)]

There was a spin-separation rocket on each RV, which was fired as the missile reached an altitude of 200,000 feet (38 miles). Its first test launch was August 1962. [Goldberg (1981)]

The A-3 has approximately 300 lbs. available for the use of decoys. ["Recommended FY 1964 - FY 1968 Strategic Retaliatory Forces," p. 23 (1962)]

Two-stage SLBM

Used energetic binder ingredient 3-nitraza-1,5-pentane diisocyanate ($C_6H_8N_4O_4$). [AD-A387 318 (1995)]

First stage:

Propellant: 20,777 lbs. of AP, Al powder, and nitroplasticized polyurethane binder.
Nozzles: Four nozzles with silver-infiltrated tungsten nozzle throat, changed from the molybdenum used in A1 and A2, for greater temperature resistance for new propellant. [Taylor (1972)] [RS (1968)] [Hove, VI-8 (1979)]
Nozzle Weight: 365 lbs each
Chamber insulation weight: 1,547 lbs.
Total first stage weight: 23,895 lbs.
Contractor: Hercules/Thiokol

Stage 2:

Propellant: 8,844 lbs. of a Hercules CMDB (HMX + NG + NC + AP + Al powder + binder) propellant
Rocket casing: Both stages were fiberglass/epoxy
Dimension: 54" dia. x 32.3' long
Weight: 35,700 lbs.

Range:  2,500 nm
TVC:  First stage changed to rotatable nozzle, and jetevator concept dropped completely, then TVC changed to Fluid Injection; Second stage TVC was based on injection of freon into the nozzles.
Second Stage Nozzles (4):  Throat material: ATJ graphite; Graphite cloth/glass-phenolic
Nozzle Weight (each):  43 lbs.;
Chamber Insulation Weight:  410 lbs.
Total second stage weight:  9,492 lbs.  [AD-A080 547 (1979)]
CEP:  1.5 nm [Caywood (1967)] [Power (1964)]
Contractor:  Hercules Inc.


Poseidon (C3) (UGM-73) (March 1971)

Main changes:  first MIRVed SLBM; diameter of missile increased 20" and 3 ft. longer for warhead payload weight increase, rather than range; first missile with single, gimbal-mounted, moveable (pivoting) nozzle replacing the previous "standard" four nozzles on each stage.  [U.S.P. #3,659,423; filed in 1964, declassified in 1972]

10 - 14 MIRV 40 kt W-68 in Mk 3 small RV's; (RV + W-68) weighs 100-150 lbs. [Francis, p. 161-2 (1995)]

Two-stage SLBM with solid fuel PCBS (Post-Control Boost System; "Bus") for MIRVs.
Dimensions:  74" dia. x 34' long
Weight:  64,400 lbs.
Missile casing:  S-901 fiberglass/epoxy, internally insulated with ethylene propylene diene monomer (EPDM).
Range:  Nominally 2,500 nm (with 10 MIRVs); 1,800 nm (with 14 MIRVs), 3,000 nm (with 6 MIRVs)
Chamber insulation weight:  2,156 lbs. (First Stage); 720 lbs. (Second Stage).
Propellant Weight:  38,800 lbs. (First Stage); 15,880 lbs. (Second Stage).
Propellant:  CMDB (Composite Modified Double-Base) (NG + NC + AP + Al + binder) formulation for first stage, Double-Base for second stage.
Propulsion System: Solid fuel; First Stage: Hercules/Thiokol; Second stage:  Hercules
TVC:  Replaced all previous 4 nozzle configurations for all stages, both stages used a single nozzle which was activated by a gas generator (which also supplied roll control); flex joint (steel reinforcements for both stages) TVC for both stages.
Nozzles (First and Second stages)
Nozzle Throat material for both stages:  Graphite-phenolic; erosion rate (at 1,000 psi pressure) = 0.0138" per sec.  [AD-A250 424 (1991)]
Nozzle weight:  765 lbs. (First Stage); 425 lbs. (Second Stage)
Total Weight:  41,860 lbs. (First Stage); 17,115 lbs. (Second Stage).  [AD-A080 547 (1979)]
**CEP:  0.28 nm** ["U.S. Strategic Objectives and Force Posture" (1971); formerly classified Top Secret]; 0.5 nm [Goldberg, part 2, p. 579 (1981)]

From 1967 until 1981, there were 41 ballistic missile submarines.

Magnetic memory cores were 170 mil tape wound, 1 mil permalloy on stainless steel bobbins. [AD-A286 599 (1972)]

When fully loaded (14 MIRV's), range of 1,800 nm.  Range increased to 2,500 nm with lower warhead loading (10 MIRV's).  Maximum range was 3,000 nm with six MIRV's.  Equipped with so many warheads because of its intended deployment in a possible Soviet ABM system environment. [Francis, p. 161 - 162 (1995)]

Range increase requirement shelved for Poseidon (confirmed by Aldridge which shows Lockheed chart with same range for Poseidon and Polaris A-3).  The throw-weight must therefore have been the principle Poseidon requirement; SLBM almost double the weight.  (so it went from 3 W-58 nukes to 10 - 14 smaller W68 nukes:  so just a guesstimate, but that would be half the weight for the W-68 compared to the W-58).

At my guess of 50% of the weight of W-58, that would make the W-68 a 100 lb. nuke, which matches previous calculation.

Greenwood, p. 169 (1988), says: 1962 LLNL started designing W-68/RV combination of 100 lbs. Mk 3 (RV + W-68 warhead) for Poseidon 160 lbs. (73 kg). Francis (1995) discloses the same time-frame of this warhead weight development (100 - 150 lb. warhead (nuke + RV)), as being for the W-68. Buchonnet also talks about 150 lb. warhead/RV's. U.S.P #4,577,812 (1986, classified 13 years) says about 150 lbs. for the warhead/RV combination with the RV being 12" in bottom dia. x 74.4" high, with a 1" dia. hemispherical top of the RV cone, and a 4.25° half-angle.

Trident I (C4) (UGM-96A) (October 1979)

A new missile and submarine approved by the Secretary of Defense in 1971 with technical specifications only agreed upon two years later, in 1973. Its principle goal was to extend missile range by 50%-100% with the same throw-weight – an ICBM range capability. [Francis, p. 163, 168 (1995)] [Hove, VI-11 (1979)] Only eight higher yield warhead RV's, lower cost devices than the previous Poseidon's. It was a move towards more accurate counter-force (ICBM targets) from counter-value (urban-industrial) targets.

Main changes: doubled range; higher tensile strength of the new Kevlar fibers than fiberglass, led to their use; pyrolytic graphite nozzle throat; addition of third stage and aerospike to compensate for the blunter nose (the latter added 300 miles to the missile range). [Spinardi, p.129 (1995)] Added a stellar INS system to the regular INS Missile Guidance System.

8 W-76 MIRV warheads in Mk 4 & Mk 4A RV. [RDD-4, V-83 (1998)]

Three stages, with the third stage through the center of the RV deck. Solid propellant PBV (Bus).
Dimensions: 74" dia. x 34.1 ft. long (the same dimensions as the Poseidon SLBM the Trident I replaced)
Weight: 65,000 lbs.
Range: 4,000 nm with a full 8 warhead payload; 6,500 nm with a lesser payload
Propellant: Composite Modified Double-Base; increased fuel and oxidizer percentage in fuel. 45% HMX (180 µ particle size), 25% NG, 7% NC, 18% Al powder, plus 5% cross-linked (long chain diisocyanate) binder. To improve the reliability of the solid propellant-to-casing adhesion, used a bonding agent (such as a brushed on 1-2 mil thick coating of undecalynic alcohol) on the uncured, 60 mil thick, elastomeric missile case-lining insulator of silica-filled, butadiene-styrene synthetic rubber. [U.S.P. #3,716,604, Hercules (1973)]
Third Stage Propellant: 20µ HMX powder coated with 1% polyureas, PEG, NG, AP, Al powder, and PEG (polyethylene glycol). [U.S.P. #5,600,088 (1997)]
Propulsion System: first stage: Thiokol; second stage: Hercules; third stage: United Technologies
Rocket Casing: The key parameters for the casing material were the specific strength and specific modulus, which should be as high as possible. Dupont Kevlar-49 fabric/epoxy was used, which allowed for higher internal pressures than fiberglass/epoxy. The adoption of Kevlar was a very important step, as it allowed much higher chamber pressures with little or no weight increase. Kevlar 49 is a light-weight polymer (1.44 g/cm3), with a tensile strength of 36 - 41 kbar with a tensile elongation of 2.8%. [Smith (1980)] [Hove, VI-11 (1979)] [U.S.P. #3,866,792 (1975)] [U.S.P. #3,843,075 (1974)] [Stellar Navigation: U.S.P. #4,306,691 (1981); 12 years classified]
Nozzle: Pyrolytic graphite plates in the throat; carbon-carbon composite (m.p. 4,000 °C) for entrance and exit segments; carbon or graphite cloth/phenolic for rest. Thermal expansion incompatibility problem improved by codepositing hafnium or zirconium carbide with the pyrolytic graphite .
Nozzle Weight: 368 lbs. (First Stage); 109 lbs. (Second Stage); Chamber Insulation Weight: 2,088 lbs. (First Stage), 913 lbs. (Second Stage), 175 lbs. (Third Stage).
Propellant Weight: 39,056 lbs. (First Stage), 17,646 lbs. (Second Stage), 3,748 lbs. (Third Stage).
Total Weight: 42,018 lbs. (First Stage), 19,034 lbs. (Second Stage), 4,070 lbs. (Third Stage).
TVC: An omnidirectional flexible joint was used for Thrust Vector Control. [AD-A080 547 (1979)]
Guidance System's Scratch Pad Memory: 2 mil plated wire memory.

After SLBM launch, an "aerospike" was extended from the nose fairing at the top of the missile to reduce boost phase drag (by 50%) due to the blunt nose fairing. Also, aluminum supporting sub-structures partially replaced with graphite-epoxy composites to give a 40% weight saving (also done with the Mk 12A RV and the ABRV).

The Post-Boost Control System has a 5,000 psia higher energy gas generator than the Poseidon, and it operated at 3,000 ºF for about 7 minutes. As a result of the higher temperature, refractory metals were used throughout. It used a dual-pressure-level, solid propellant (85% HMX, 15% HTPD/IPDI) gas generator system. It also used niobium (89% niobium, plus 10% hafnium, and 1% titanium) or tantalum (90% tantalum with 10% tungsten) alloys with a coating of hafnium disilicide (HfSi2) for protection against the oxidizing propellant thrust gases.

Two Trident submarine U.S. support bases located at Kings Bay, Georgia and Bangor, Washington.

Trident II (D5) (UGM-133) (1990)

8 W88 800 kt warheads in Mk 5 REB MIRVs when fully loaded. (Also can be loaded with 12 100 kt W-76 MIRVs, instead of 8 W88's.) The Mk 5 REB heatshield is made of tape-wrapped, rayon fiber-based carbon-phenolic composite. [AD-A318 763 (1996)] Fourteen Ohio class submarines (powered by an S-8G nuclear reactor) with 24 Trident II missiles in each one.

Like Trident I, a three-stage missile, with third stage through the RV deck, plus the PBV.

Dimensions: 83" dia. x 44.5" high
Weight: 126,000 lbs. (57,272 kg)
Range: 4,230 nm with 8 RV's; 6,000 nm at reduced load. At a range of 5,200 nm, the apogee is 825 nm above the earth. [AD-A272 447 (1993)]
Propellant: Nitrate ester (NG) plasticized with PEG (polyethylene glycol). MTN (metrioltrinitrate) is used in conjunction with the polymeric binders. 60 μm or less spherical aluminum powder. MTN mixed with 8% metriol triacetate it gelatinizes/plasticizes NC. [AD-A387 318 (1995)]
First and second stage: Hercules/Morton-Thiokol.
Third Stage: UTC (United Technologies Corp.)
Missile Casing: 42" wide woven graphite fiber cloth impregnated with epoxy for first two stages; Kevlar-49-epoxy third stage
Nozzle: Carbon-carbon composite
Fore and Aft casing insulation: EPDM (ethylene propylene diene monomer, a rubber compound is applied, and vulcanized.
Guidance: INS and Stellar

Manufacturer: Like all six FBM's (Fleet Ballistic Missiles), the Lockheed Missile & Space Co.

[Cochran (1984)] [Spinardi, p. 159 (1994)] [Dalgleish (1984)] [Drell, p. 28 (1991)] [U.S.P. #4,711,086 (1987)] [AD-A511 972 (2009)] [U.S.P. #5,413,859 (1995)] [U.S.P. #4,654,103 (1987)] [GE Made RV: U.S.P. #4,892,783] [U.S.P. #4,623,106 (1986)]

| | RV's/Warhead | Throw-weight | RV + nuke (estimated) | Yield |
|---|---|---|---|---|
| Polaris A-3 | 3 x Mk 2/W-58 | 400 kg | 136 kg | 200 kt |

| | | | | |
|---|---|---|---|---|
| MM II | 1 W-56 | 500 kg | 500 kg | 1.27 MT |
| MM III | 3 x Mk 12A/W78 | 1,100 kg | 365 kg | 335 kt |
| Poseidon | 14 x Mk 3/W68 | 2,000 kg | 75 kg | 40 kt |
| Trident I (C4) | 8 x Mk 4/W76 | 1,500 kg | 190 kg | 100 kt |
| Trident II (D5) | 8 x Mk 5/W88 | 2,800 kg | 350 kg | 800 kt |
| MX/Peacekeeper | 10 x Mk 21/W87 Avco Corp. RV | 3,950 kg | 395 kg | 335 kt |

[PBV for 3 MIRV MM III weighs 1,000 kg [AD-B060 927 (1981)]] [Gertz, p. 252 (1999)] [Throw-weights from Harvey (1994) and Nitze (1979) for MM III] [Buchonnet, p. 38 (1976)] [Drell, p. 40-41 (1983)]


**Other Nuclear-Armed Missiles**


Terrier tactical missile (RIM-2E) (Naval SAM) (1955)

W-45 fission warhead

Two-stage, medium range defensive SAM.
Dimensions:  18" dia. x 27' long
Launch Weight:  3,000 lbs.
Range:  Initially 20 miles, upgraded to 40 miles.
Solid Propellant Booster and Sustainer
Booster Motor-Case Material:  AISI 4130, 4140 steel; case length:  131.26"; wall thickness:  0.087 - 0.099"
Sustainer Motor-Case Material:  AISI 4130, 4135 steel; case dimensions:  13.5" dia. x 55.84" long; wall thickness:  0.063 – 0.068"
Yield Strength for both booster and sustainer casing:  75,000 psi;  tensile strength for both:  95,000 psi

First deployed in 1955 on the first guided missile heavy cruisers (cruisers are one down in size from battleships), the Baltimore class.

Prime Contractor:  General Dynamics/Pomona


Genie Air-to-Air Missile (MB-1) (January 1957)

Warhead:  1.7 kt W-25 pure fission, composite core, 17.4" dia., 218 lbs.

Fuel: Ammonium Perchlorate
Rocket Liner:  Polyhydrocarbons polybutadiene rubber  [AD-A210 747 (1989)]

At 10,000' the 2 kt radiation envelope of 5000 R (immediate kill) has a greater range than thermal and wind effects.  At 40,000' radiation the greatest range of all effects. ["GACAEC", 41st, p. 7 (July 1954)]

TALOS-W (RIM-8G) Naval SAM (1957 – 1979)

W-30Mod2 22" diameter, boosted Oralloy fission or conventional HE warhead

Two-stage; 40,000 hp ramjet-powered missile with Mach 2.5 speed
Range:  2 - 80 nm
Velocity:  2,000 ft./sec
Dimensions:  30" dia. x 38' high with booster; 32' w/o booster; 9' wide wing span for stabilizing wing
Weight:  3,400 lbs.; 7,700 lbs. with booster   [RS3434/15 (1967)]
Booster Propellant Type:  Solid double-base
Sustainer Propellant Type:  Bendix Ramjet JP-5
Motor Case Material:  AISI 4130, 4140 steel; stamped and drawn or welded; wall thickness 0.150 - 0.165"
Case Material Yield Strength:  160,000 psi minimum; Tensile Strength:  185,000 psi minimum [AD818 416 (1967)]

Contractor:  Bendix Corp.

Retired from 1974 to 1979.  Deployed as a defensive Naval radar-guided, beamriding SAM on surface ships, but could also be used as a SSM.  HE warhead deployed/used in Vietnam War.  U.S.P. #3,908,933 (1975), "Guided Missile", is a 109 page, detailed patent of just the missile itself.

Nike-Hercules SAM (MIM-14B) (mid-1958 – 1979)

30" dia. boosted W-31; 3 separate warhead yields; additionally, the Nike B has surface-to-surface, rather than surface-to-air, 2 kt or 30 kt yield [SC-TM-188-56-51 (1956)]

Originally called the "Nike I."  In addition to the Sergeant, several other motors employed polysulfide rubber binders. The sustainer motor of the Nike-Hercules surface to air missile used Thiokol rubber, as did numerous small sounding rockets, and the retro-rockets of the Mercury and Gemini manned spacecraft. 2,520 built and fielded, most of the improved MIM-14B.

Dimensions:  34.6" dia. x 41' 6" high; with the booster 43.1" in dia. and 16' 11.5" long, and the sustainer
             36" in dia.
Propulsion:  Quad solid-fueled rockets; Thiokol M30 with Hercules M42 booster; all solid fuel rockets
Propellant:  Nike:  NC + NG + Triacetin with Flameristic as the liner;  Nike Sustainer:  AP + polysulfide,
             with polysulfide rubber liner  [ AD-A210 747 (1989)]
Weight:  10,710 lbs.
Range;  93 miles
Speed:  Mach 3.65
Altitude:  100,000'
Guidance:  Radar command
CEP:  150 m.  Has an accurate surface-to-surface mode [LA-4350-MS (1969)]

Contractor:  Western Electric/Douglas Aircraft for the U.S. Army

Missile propellant for all three stages (booster, sustainer, and jethead) is polysulfide.  Initially, has a solid rocket booster that provides thrust for 5 seconds, and then separates shortly thereafter.  A sustainer solid propellant rocket stage then fires and increases missile velocity to 11,400 ft./sec. its maximum, at burnout within about 12 more seconds.

The gross weight is 20,000 lbs. at takeoff, with the booster weighing 11,761 lbs., and 3,700 lbs. at burnout.

ASROC (RUR-5A) (1959)

W44 fission warhead

Single stage, anti-Submarine ROCket, solid-fueled, unguided ASW surface-to-underwater depth charge missile; believed to be two-point initiation

Propellant:  Solid, extruded double-base
Motor Case:  Steel, rolled and welded; case length:  56.3", wall thickness 0.2"; yield strength:  160,000 lbs. minimum [AD818 416 (1967)]
Range:  1 - 6 miles
Depth Charge Configuration:  13" dia. x 155" long
Torpedo Configuration:  13" dia. x 180" long
Launch Weight:  940 lbs.  [AD818 416 (1967)]

Contractor:  Honeywell Inc.  [Samuelson (1967)]  [AD-A995 502 (1963)]

Falcon (AIM-47A) (GAR-11) Air-to-Air Missile (1962)

W-54 fission warhead

Dimensions:  11.4" in diameter
Weight:  260 lbs. (200 lbs. of fuel)
Speed:  Mach 6
Range:  100 miles
Total Impulse:  12,900 lbs.-sec.
Propelled by a one-stage Thiokol M60 solid-fueled rocket motor.
Guidance:  IR/pulse Doppler radar  [RS 3434/35 (1968)]  [Taylor (1972)]

Manufacturer:  Hughes

Bullpup B (AGM-12D, originally GAM-83B) (1959)

W-45 fission warhead

Air-to-surface, single stage, liquid-fueled Navy/Air Force tactical missile; also SAM (surface-to-air missile)

Dimensions:  18" x 13'7"
Weight:  1,785 lbs. (810 kg)
Range:  9 nm
Guidance:  Radio command
Powered by Thiokol LR-62 storable liquid propellant rocket motor
Oxidizer:  IRFNA (Inhibited Red Fuming Nitric Acid), Type III-A (14% $N_2O_4$ + 0.6% HF) (83.5 lbs.)
Fuel:  MAF-3A (Mixed Amine Fuel) (~20% by weight UMDH, ~80% DETA (diethylenetriamine), and acetonitrile) (also – like oxidizer – inhibited Type III-A) (28.6 lbs.)
Both fuel and oxidizer tanks were made of 2014-T6 aluminum

Contractor:  Martin Marietta

[Clark (1972)]  [UCRL-ID-125506 (1962)]  [Taylor (1972)]  [AD857 651 (1969)]  [AD818 416 (1967)]

Bomarc (CIM-10B) (Bomarc A: 1959; Bomarc B: 1961)

W-40 warhead

Dimensions:  35" x 45' 1" with booster; Wingspan:  18' 2"
Weight:  16,032 lbs.
Speed:  Mach 2.8
Range:  440 miles (700 km); Effective height range:  100,000'
Booster Fuel/Oxidizer:   Bomarc A:   JP-4 and Unsymmetrical Dimethyl Hydrazine; Bomarc B:   solid propellant
Propulsion:  Bomarc B:  two Marquardt RJ43-MA-7 ramjet engines with 12,000 lbs. thrust, with a Thiokol M-51 solid-fueled (Bomarc B) integral rocket booster of 50,000 lb.f thrust for 30 seconds.

Contractor:  Boeing Co.

In 1961, there were 238 Bomarc's.  383 Bomarc's were the peak number in 1963.  Phased out of service in 1972.

The Bomarc was a winged, ramjet-powered surface-to-air defense missile.  It was command guided to the target, with active homing during final approach to the target.

SUBROC (UUM-44A) (1965-1989)

W-55 warhead

SUBmarine ROCket, ASW; single stage, submarine-launched; surfaces to air then plunges to underwater missile for final targeting sequence.  [U.S.P. #3,853,081 (1974), classified 16 years)]

Propellant:  solid composite polyurethane
Propulsion System:  Thiokol TE260 solid-fuel rocket
Range:  70,000 yds.  [AD-A036 610 (1964)]
Dimensions:  21" dia. x 22' long x 0.219" wall thickness
Case Material:  AISI 4132 steel  [AD818 416 (1967)]
Weight:  4,000 lbs. (1,815 kg)
Inertial-guidance by General Precision Inc.

Contractor:  Goodyear Aerospace Corp.; Motor contractor:  Thiokol.  Built for the U.S. Navy.

SRAM (AGM-69A) (1972) (Built FY1969 - 1973)

W-69 warhead

Short-Range Attack Missile; air-to-ground B-52 missile for Soviet air defense suppression or hard target kills; a stand-off, long-range attack weapon; the SRAM was originally targeted primarily for penetration, so that the bomber can drop some of the larger bombs (like the B-53).  It became a stand-off weapon, where the B-52 could remain safely outside of the enemy border, while firing its SRAMs into there, and the SRAM had a small radar cross section in aid of its mission.  [AD-A032 600 (1976)]

The SRAM could be carried by the B-52 or the FB-111.  The FB-111 could carry 4 SRAMs mounted on its wing pylons.  The SRAM was a replacement for the two Hound Dog turbojet engine that were carried under the B-52's wings.

Dimensions:  17.5" dia. x 14'
Weight:  2,240 lbs. (1,015 kg)
Casing:  graphite fiber/epoxy
Nose Cone:  Possibly 10° half-angle, ½" diameter, boron-doped (0.2 - 1.6 weight %) pyrolytic graphite
        with a nose thickness of 1" over a 5" tall hollow conical base (the nose tip's "skirt") of diameter
        2.54" and thickness of 0.18"; nose thickness to skirt thickness ratio of 5:1 [AD824 697 (1967)]
Propellant:  dual thrust, concentric, HTPB-bound grain; initial ignition after dropping 2,000 feet below B-
        52.
Speed:  Mach 4 ["JCAE", Part1, p. 9 (1973)]
Boost phase:  20 - 40 seconds at 2 g's:  400 - 800 m/s
Range:  30 nm at low altitude; 70 nm at higher altitudes

The SRAM solid rocket motor was 100" long, 17.6" in diameter.  It contained two end burning solid propellant grains with a total propellant weight of 1,000 lbs.  A composite propellant is used with a burn rate of 2.3" per second. [AD-A014 428 (1975)]  Uses a high pressure (3,000 psig) (pressurized total weight: 9.3 - 9.4 lbs.) helium gas bottle to provide an initial launch boost to the missile's hydraulic flight control system.  The hydraulic system accumulator/regulator is pressurized by the helium.   [AD-A208 994 (1989)]

Contractor:  Prime contractor was Boeing Co.; also Aerojet-General, and Lockheed (contractor for the
        rocket motor)

Carried by B-52 on an internal 8 round rotary "revolver"-type rack; and 12 carried externally on the wings; thus a total of 20 SRAMs can be carried, plus the regular bomb load of four Mk 28 bombs (later the B61).  The SRAM essentially MIRVed the B-52 Bomber, thus keeping the B-52 in the Nuclear Triad, and not be retired as obsolete.

Full production of SRAMs was 40 missiles per month.  1,451 SRAMs were in service at their peak numbers in 1975.  After 1972, a sample bomb load was 8 SRAMs and a random mix of 5 aerial bombs (Mk 28, Mk 43, Mk 57, B61, or a single B-53 alone).  [AD-A238 593 (1985)]

[Orbital Launch Systems Group web site (2004)] [LA-11401, p. 21 (1991)] [Dorman (1995)] [ADC 021 800, p. 48, 52 (1980)]


**Anti-Ballistic Missiles (ABM)**


Sprint ABM Missile (1974)

W-66 2 kt ERW warhead; payload is 40 - 100 kg (typical payload 40 kg) [AD-A345795 (1992)]; though mainly a neutron warhead, its prompt gamma ray output was almost 7 times that of an ordinary fission weapon.  Its gamma ray output was $1.3 \times 10^{24}$ MeV = $5.0 \times 10^7$ kcal, with an average energy of 2 MeV, compared to 1.5 MeV for a standard fission weapon.  [Northrop (1996)]

Conical-shaped, two-stage SAM Missile.  It was essentially a missile shaped like a giant conical-shaped MIRV to reduce aerodynamic heating, allowing the designers to maximize the missile's speed.  It was fired with a 25° launch-angle trajectory.  [AD-A033 540 (1976)]

Range: 15 - 25 miles
Altitude: 5,000 - 100,000'
Dimensions: Total: 54" base dia. x 27' high; 2nd stage: 36" base dia. x 8'3" high. At the top of the first stage was an internal 51" diameter 2014-0 aluminum dome. The monocoque airframe used fiberglass filament-wound motor cases as a primary structure. The remaining load-carrying structure was aluminum.
Weight: 7,500 lbs. (3,425 kg) [Burnett (1992)]
Contractor: Both stages by Hercules

Velocity: (From 20 - 70 km) 10,000 feet/s [AD-A955 400 (1972)]
First Stage Propellant Weight: 675 kg in a 6-star point configuration [AD-A345 795 (1992)]
First Stage Acceleration: 130 g's
Second Stage Acceleration: 360 g's; highest acceleration of any U.S. missile; used a steel motor housing
High Acceleration Propellant Configuration: U.S. Patent #3,664,133 (1965).
Second Stage Insulation: The thermal protective system of the second stage is designed to maintain the structural integrity of the aluminum shell and the fiberglass motor case. The ablative nose cone consists of a hemispherically tipped, ½" diameter nose cap flaring into a 6° half-angle conical section. The nose cap is formed with a center rod of quartz phenolic over which is wrapped phenolic-impregnated silica (quartz) tape. The ablative shielding covering the rest of the second stage is fabricated from silica cloth impregnated with phenolic resin mixed with rubber for additional elasticity. The leading edges of the air vanes is protected by molded edge-oriented quartz-phenolic tape. [AD-A004 431 (1975)]
Guidance: Western Electric (missile guidance set and accelerometers) and Honeywell (gyros, etc.)
TVC (Thrust Vector Control): By injection of liquid Freon as a vectoring fluid to provide pitch and yaw forces, through four 3-barrelled injection valves into the Sprint's propulsion nozzle.

Contractor: Martin Marietta, Orlando Division

Outgrowth of Nike-Zeus (renamed Nike-X) research. Nike-Zeus technology "flawed and unstable". Research until 1968.

Sphere-cone nose tip of porous, liquid Freon transpiration-cooled tungsten, on a base of 90 weight % tungsten, 10% copper. [AMMRC CTR 76-38 (1976)] [U.S.P. #3,883,096 (1975)] The two nozzle exit cones were made of copper-infiltrated tungsten. The nose tip had a symmetrical ablation rate of .34 inches per second. [AD-A033 540 (1976)] The transpiration mechanism was fed from a conical reservoir, which was found to be mandatory from a volumetric efficiency viewpoint. With the conical reservoir, a bladder expulsion technique must be used. The design coolant weight was 1 - 2.3 lbs. of liquid, expelled out of the silicon rubber bladder with possibly a cold gas pressurization system using high pressure helium contained in a spherical pressure vessel, pressurized to 7,000 psi. [AD767 570 (1972)]

The superstructure is of aluminum coated with 60 mils of epoxy-phenolic film adhesive and one ply of silica fabric-phenolic resin. [U.S.P. #3,830,666 (1974), Army]

Ceramic was used in its high voltage capacitors. The CDU (Capacitor Discharge Unit) used mica paper or bentonite as the dielectric in its high voltage capacitors and. [AD-A286 599 (1972)]

Sprint required a high burning rate solid propellant an order of magnitude higher than that in use in such missiles as Polaris, Minuteman, and the Pershing.

The FAE-7 Hercules composite double-base propellant used in the Sprint was made up of (weight percent):

36% ultra-fine ground ammonium perchlorate of 9-10 micron (μm) diameter particle size,
30% nitroglycerin,
18% nitrocellulose,
7.2% aluminum powder, chopped Al foil, and Al staples, and

6.7% triacetin, 1.1% resorcinol, and 1% 2-nitro-diphenylamine

The propellant had a density of 107 lbs./cu.ft., a burn rate of 3.6" per second, and a vacuum specific impulse of 305.3 sec. [AAMRC CTR 74-47 (1974)] The exhaust product contains 13.4% $Al_2O_3$ (aluminum oxide) particles. Combustion chamber temperature was 6,037 °F. The nozzle was 80% tungsten, impregnated with 20% copper, making it easily machinable.

In comparison, there was also CMDB (Composite Modified Double-Base) propellant stabilized with urethane compounds: 41% HMX, 20% Al (30 µinch), 19% NG, 8.2% NC, 8% AP (U.S.P. #4,478,956, filed 1973, declassified 1984; Hercules).

Spartan ABM Missile (LIM-49A) (1974)

4 MT W-71 warhead

Three Stage, Solid-Fueled, SAM Ballistic Missile
Dimensions: 42" dia. x 55' long
Total Rocket Weight: 15,000 kg [AD-A288634 (1994)]
Speed: Mach 10 (7,418 mph)
Propulsion: Thiokol TX-135, TX-238, and 3rd stage, TX-239 (2.65' dia. x 8.5' high; 0.8 m x 2.6 m) motors; second stage rocket case made of ultra-high strength 4340 steel alloy [AD-A066814 (1979)]; first, second, and third stage domes of thin gauge 4340 steel. [AD748416 (1972)] The rocket motor was also made of 4340 steel. [AD-A066814 (1979)] The first stage was identical to the Nike-Zeus, which became the second stage of the new, enlarged Spartan.
Operating Range: 70 - 100 miles high (160 km) [AD-A338560 (1995)], 300 nm [Goldberg, p. 556 (1981)]; 4,000 cubic miles (up to a distance of about 16 miles) destructive range for hardened RV's, and 4 million cubic miles (up to a distance of about 160 miles) for unhardened RV's.
Guidance System Weight: 90 lbs. designed by Bell Telephone Labs

Contractor: Western Electric Co. Inc. and McDonnell Douglas

The Spartan missile was a direct outgrowth of Nike-Zeus and Nike-X research. It was a longer range missile than its predecessor, the Zeus, with a range of 300 nm.

The Spartan SAM which carried the W-71 was a short three-stage missile, only 16.83 m (55' 2") long, with all three-stages manufactured by Thiokol. The missile diameter was 43.1", and the warhead end tapers to probably less than 36" in diameter. It had a 51" diameter 2014-0 Aluminum spherical dome formed initially by spinning, and then to save money used explosive forming [AD748 416, "Center for High Energy Forming – Final Report" (1972)]

It was a fraction of the size of an ICBM, yet it boosted its 4 MT warhead into the exo-atmosphere (ceiling 350 miles/560 km), at high speed. It was designed to exit the atmosphere and its x-ray flux would disable incoming RV's. [AD-A241725 (1991)]

The nuclear warhead is specially wrapped with a light weight insulating/shielding layer to protect it from nuclear blasts. It consists of a 0.125" aluminum sub-layer, bonded with a 54 mil tin sheet over-layer. The tin layer is over-wrapped with three or four tape layers of epoxy-glass filament spray-coated on top with a thermal insulating ablative layer consisting of an epoxy resin mixed with silica fibers, Cab-O-Sil colloidal pyrogenic silica pigment, and 22 - 23 wt.% subliming salts, including di-ammonium borate. [U.S.P. #4,041,872 (1977), classified 6 years; Army]

Ceramic was used in its high voltage capacitors. The CDU (Capacitor Discharge Unit) used mica as the dielectric in its high voltage capacitors. [AD-A286 599 (1972)]

The ground real-time Data Processing Unit had to handle up to 10 million instructions per second, and a peak I/O transmission for radar control of about 70,000 64-bit words per second, about three times the throughput of a standard American 1.544 Mb T1 (24 channel telephone) trunk line. A multiprocessor with the capability of 10 microprocessors was used.

The operating system had 750,000 instructions, with a total of 2,100,000 instructions for the entire system. It was written in assembly language, PL/1, and FORTRAN. It was contained in 50 digital racks, each containing up to 100 logic chassis, each containing 500 - 600 logic gates.

Walleye (AGM-62A) Air-to-Ground Glide Bomb (1967)

W-72 low yield warhead

A glide bomb with no propulsion. The pilot captures the target on a TV camera, and the image guides the glide bomb to the target with perfect accuracy, similar to the Condor bombing system.

Dimensions: 15" x 11' 3"
Span (fins): 3' 9.5" (1.56 m)
Weight: 1,125 lbs. (510 kg)
Range: 19 miles (30 km)
CEP: 0

Contractor: Martin Marietta

Non-nuclear armed version used by the U.S. against bridges and such in the Vietnam War, though too late to have an effect on the outcome of the War.

An ECCM version is described in U.S. Patent #5,004,185 (issued in 1991, filed in 1964; 27 years classified).

# Stockpile-to-Target Sequence

## Reentry Vehicles and Ballistic Missile Delivery Accuracy

"It has always been my opinion that when a man sets
himself determinedly to do something and thinks of
naught but his design, he must succeed despite all
difficulties in his path..."

-- "The Memoirs of Jacques Casanova
de Seingalt" (1774)
Giacomo Casanova

"The Day of the Son of Man will be like lightning that
flashes from one end of the sky to the other."

-- Luke 17:24

The electronics revolution. Before the I.C. (Integrated Circuit chip), before the transistor, first there was the glass "vacuum tube", invented in the 1930's by Lee DeForest. It was large, fragile, expensive, and produced a lot of heat, thus requiring fans and ventilation. But it made possible the electronic digital computer. But it was huge. The first computer, named ENIAC, was built for the U.S. military in 1944. It used 18,000 vacuum tubes, weighed 27 tons, consumed 200 kW of power and had a floor space of about 12 meters by 12 meters (40' by 40').

Then in 1947, there came the invention of the tiny germanium (now silicon) transistor that replaced the tube.

The transistor had been an individual part – a "discrete" component – that needed to be wired together with other components – resistors, capacitors, inductors, etc. – to make a functioning circuit. A big improvement in size, cost, and power consumption, but still a huge wiring and soldering problem.

The wiring problem was solved in the 1940's by the printed circuit board (PCB), a thin insulating fiberglass-epoxy sheet, laminated on top with a thin layer of conducting copper metal. The copper was mostly etched away – printed – in places to replace the previous tangle of wiring, and the components soldered to the copper film. First there were single-sided boards, then double-sided boards with copper films on both sides of the fiberglass-epoxy sheet, and finally multi-layer boards. Each an improvement, and a further step in size/volume reduction. You can see PCB's in every modern electronic device with its cover off.

Then came the final improvement: the integrated circuit – the "IC", or chip. The IC combined and miniaturized numerous transistors, capacitors, and other components into a single, very small plastic-encapsulated package. Jack Kilby at TI (Texas Instruments) and Robert Noyce at Fairchild Semiconductors are credited with their simultaneous discovery of the IC, both filing independent (but competing) patents for it in 1959.

The manufacture of the IC produces a much more compact device, but uses a much more complex process to manufacture, with multiple layers on its thin, flat, ceramic "board", on which are "printed" the transistors, resistors and capacitors, and thin metallic conducting layer that connects these components, and the thin, insulating ceramic board to keep them electrically isolated (originally it was a thin layer of silicon that formed the chip).

In 1961, the first IC computer was developed, the Solid Circuit Network Computer, weighing 300 g/10.6 oz. – compared to the same tube computer that weighed in at 13 kg (28.6 lbs.) – which was used for the Minuteman I program, whose first missile became operational in 1962.

## Reentry Vehicles

What goes up, must come down.

A ballistic missile, either an ICBM or SLBM, has a point programmed into it.  This is the target.  The missile's Reentry Vehicle coasts across the sky aimed at the target, and hits near the target.  The last chance for target coast correction is during the initial boost phase, fairly early in flight, when the engines run out of fuel or are shut-off, and the post-boost shutoff phase begins.

This is programmed for a certain point in 3-D space with a certain speed and direction for the coasting phase to the hit the target.  This is necessary due to variations in engine thrust, missile weight, missile aerodynamics, winds, atmospheric conditions, and gravitational variances, whose effect on trajectory is magnified by the intercontinental distances required for the ICBM to traverse.

When all of the three stages of a modern strategic solid propellant ICBM each burn out in a couple of minutes or so, and fall back to earth, the "boost phase" is at an end, and the reentry vehicles mounted on a "bus" continue by momentum on unpowered flight, coasting along on an elliptical path towards their designated targets.  At previously specified points, in a multiple RV ICBM, the individual MIRVs are ejected from the bus towards their separate targets.

When they realized that atmospheric reentry was difficult, they set to work designing the perfect "Reentry Vehicle", or RV, to survive the tremendous heating of atmospheric reentry, and then once that problem had been solved, carry the nuclear warhead as accurately as possible to a point in the atmosphere above its DGZ (Designated Ground Zero) target thousands of kilometers away from its launch point.

It turned out to be a tremendous problem – the most difficult of all, in fact – in designing the ICBM/SLBM.

Because actual testing of an RV was extremely expensive, difficult, and specialized, wind tunnels for high Mach speed air flow had to be designed and built.  The technology achieved almost full maturity in the relatively short period of time of about a decade, from 1959 - 1970.  From the 1959 Thor IRBM's GE-manufactured Mk 2, a 3,500 lb., **blunt**, beryllium, **heat sink** RV, to the 1970 Minuteman III's Mk 12,125 kg, **conical**, carbon-carbon composite, **ablative** RV.

From crude and inefficient blunt RV's to the much superior, light weight, high speed, and highly accurate, cone-shaped RV's.

The ICBM carries the nuclear warhead in a Reentry Vehicle, or RV, whose dual purpose is to:

1)      protect the nuclear weapon from the high heat generated by its rapid descent through the atmosphere, and

2)      accurately maintain its ballistic trajectory, after being guided by the missile's INS (Inertial Navigation System), to hit as close as possible to the target point the missile has been aimed it at.

From 10,000 km away, to hit within a target area sized an area of a couple of football fields was the pinnacle of 20th Century technological supremacy. And the U.S. spent a lot of effort, and 100's of millions of dollars accomplishing it. As with the development of the rocket itself, and its solid-fuel propellants, RV's have received a huge amount of research attention, as evidenced by the amount of research dollars spent, and the number of patents issued for each.

With the introduction of MIRV's in 1970, each ICBM releases its multiple, conical RV's (actually with a hemispherical nose-tipped cone), spinning at a moderate speed (on the order of one rotation per second) on its longitudinal axis for stability, for accuracy, and even reentry heating, a nuclear weapon-bearing MIRV reentry vehicle has to descend back through the atmosphere towards its target.

Starting at around an altitude of 90 - 100 km, accelerated by g, the 9.8 $m/s^2$ acceleration due to the earth's gravity, the RV reaches a speed of around 11 km/s or about Mach 35, and in doing so the atmospheric friction becomes a source of tremendous heat. Peak heating occurs near 20 km, and peak deceleration is ~50 g. The maximum temperature reached is ~7,200 °C, and peak heating occurs at an altitude of 15 – 50 km. [AD-A224 584, p. 2-25 & 3-40 (1990)] This heat must be anticipated and dealt with, or the RV will burn up in the atmosphere before reaching its target.

> "A ballistic missile follows an elliptical flight path. For a given initial velocity, there is an initial flight path angle that results in maximum range, and the trajectory is called the minimum energy trajectory. For minimum energy trajectories, the flight path angle decreases from 45° for very short range to approximately 32° [angles are all from the flight path with the horizon] for a range of 3,000 nm; the velocity increases…to approximately [7 km/s] …[and, finally] a maximum apogee of about 700 nm occurs for a range of 5,500 nm and a flight path angle of 22.5°…
>
> Trajectories for which the initial flight path angles are greater than the minimum energy trajectory are called lofted trajectories…" [Glover, p. 79 (1971)]

For a "lofted trajectory" – the closer to the vertical that the RV trajectory takes it – the faster the speed of reentry, the more accurate is the RV, and the more immune it is to enemy counter-measures. But there is a price to entry at a high angle, other than the increased missile energy, and thus the resulting maximum RV height above Earth obtained. It's the higher speed of reentry.

During the RV's high speed descent, its kinetic energy and the resulting atmospheric friction generated is converted to an enormous amount of heat, which the RV and its contents must dissipate if it is to survive and arrive intact. The amount of heat involved is around 100 - 150 kW/cm2 or 5 kcal/g.s of RV, high enough to vaporize carbon, the element with the highest boiling point (4,827 °C/8,721 °F; actually carbon doesn't melt – which is even better – but sublimes directly from solid to vapor). But at least lofted trajectory atmospheric reentry only takes a reduced time of about 60 seconds of this heat.

The only problem was that pure carbon, by itself, is not a suitable material to make RV's out of, by virtue of the fact that its physical properties preclude such a use. Carbon (or its allotrope graphite) is a crumbly powder that cannot physically stand up to any sort of mechanical stress – it simply falls apart.

Yet, after about a decade of intense research, experimentation, and testing, they eventually did the job with a basically conical-shaped RV with a surface layer only 0.5" thick of carbon-carbon composite, a dense pyrolytic carbon composite first fielded in the Minuteman III Mk 12, and then the Mk 12A RV's.

The carbon-carbon composite is manufactured starting with a carbon felt or graphite fiber weave heated to 1,100 °C in a methane atmosphere to decompose the methane and deposit the pyrolytic graphite which fills in all the holes and gaps in the carbonized viscose rayon felt (of 23 micron (µm) fiber diameter) or graphite fiber weave. [SC-DR-69-126 (1969)]

The reentry heat is primarily developed and largely concentrated in the nose tip of the RV, which is an ablative, sacrificial item as a result, and is hemispherical in shape, rather than pointed, to minimize the shock-heated air and the resulting increased nose tip heating and ablation. The hemispherical shape accomplishes this by minimizing the "bow shock", and thus the heating of the nose tip.

The effect of reentry is thus mostly based on the shape and material of the nose tip. Pyrolytic carbon is one of the best heat conductors of all in the lateral/horizontal direction (the C direction of the A-B plane, aka the Z direction of the X-Y plane), along with a high melting point, high heat absorbing capacity, and it thus conducts the heat away from the nose tip, downwards to the rest of the carbon coating of the MIRV. Thus, the heat is spread out, rather than concentrated in the nose tip, to be diluted and radiated off a much larger surface area.

A third source of RV thermal radiation are the ablation products injected into the thin protective air boundary layer that naturally surrounds the descending RV body. Only certain purely organic heat shields produce a char of sufficient strength to withstand the severe shear stress imposed during atmospheric reentry. On the other hand, silica-phenolic heat shields retain their outermost layers, introducing primarily gases (CO, CO2, and H2O) into the boundary layer, which produce intense heat radiation, further cooling the RV.

In spite of all this, the RV graphite surface usually reaches a temperature in excess of 2,000 °C. [AD763 495, p. 2-119 (1971)]

The overall heat transfer balance at hypersonic velocities consists of:


Heat In = Heat Stored + Heat Out

= Heat Storage Rate + (Surface Vaporization + Radiative Emission + Heat Conduction Downward from the nose tip all the way to the bottom of the RV)


At 3,400 °C., just below the sublimation temperature of graphite, the cooling radiative emission of the descending – now glowing incandescent – RV is 850 J/cm2.s = 200 cal/cm2.s (by the Stefan-Boltzmann Law, the radiated $E = \sigma T^4$, where T is in degrees Kelvin, and $\sigma = 8.9 \times 10^{-12}$ J/cm$^2$.K$^4$.s).

The surface area of the Mk 12 RV is about 230,000 cm$^2$, which gives a radiative energy of 200 MJ/s.

With a ballistic missile with a nominal range of 5,500 nm (6,600 miles/10,600 km) the angle of 10° trajectory (the reentry angle is defined as the angle the RV path makes with the horizon) is 24,300 ft./s, while at an angle of descent of 60°, the speed is 31,600 ft/s (10 km/s; Mach 30) (30% higher). The peak heating occurs at an altitude of 20 km. [AD405 855 (1963)]

At a 10,000 km-range, the ICBM begins its entry into the atmosphere at an altitude of 150 km. The horizontal range of the missile from reentry (height = 150 km) to impact is 330 km, and the time to impact is about 65 seconds. On the other hand, the earlier minimum energy trajectory has a reentry angle of 24°, and peak heating occurs at 23 km, with a peak deceleration of 50 - 60 g's, and the speed of reentry is around 7 km/s (around Mach 20). ["Countermeasures" (ca. 2000)] [AD-A229 778 (1990)] A 15° reentry angle is a depressed trajectory results in lower accuracy, but makes the RV harder to detect because of its lower maximum height, but requires more insulation to overcome atmospheric frictional heating.

A lofted trajectory is a 35 - 40° reentry angle, and requires more energy to send the ICBM higher above the earth's surface. [AD-A229 778 (1990)] [AD-A398 731 (1963)] For a 40° reentry angle, the speed of reentry is around 8 km/s. [AD379 893 (1967)]

The key parameter governing reentry heating is the RV's ballistic coefficient, β (Beta: the weight-to-drag ratio), which is given by:

$$\beta = W/C_D A$$

with a higher β value resulting in a faster descent, and where W is the weight of the RV, A is the maximum cross-sectional area perpendicular to the direction of motion (the conical RV's base area), and $C_D$ is its drag coefficient. ($C_D = 2 \sin^2 \theta$, for a cone of half-angle θ°) The drag coefficient has values around 0.03 - 0.05 for an aerodynamically streamlined body, and 1.0 – 1.5 for a non-streamlined one. [Ivanov, p. 258 (1989)] The higher the value of the ballistic coefficient, the less the object is slowed by air resistance and the faster it falls through the atmosphere. It is a factor in the promulgation of laminar flow, as opposed to turbulent flow, that slows and heats the RV. It is the sharp drop in velocity that is responsible for the emergence of eddies (turbulization) in the air flow.

Modern cone-shaped RV's have values of Beta in the range of 100,000 - 150,000 N/m2 (2,000-3,000 lb/ft2). An RV with a β of 150,000 N/m2 has a maximum heating rate of 2,500 cal/cm2.s, and would reach the ground (from the upper atmosphere) in 55 seconds. β numbers above 1,000 lbs. per sq.ft. (first achieved with Minuteman II ICBM's RV) travel through the atmosphere with sufficient speed to remove much of the contribution to inaccuracy of the reentry phase to the CEP. [Goldberg, part 2, p. 586 (1981)]

$$P = (\sin \alpha_E) \times \beta$$

Where P is the atmospheric pressure at peak deceleration, and $\alpha_E$ is the reentry path angle (with the horizontal plane) at 300,000 feet. [AD353 247 (1964)]

Essentially, the problem of over-heating for the RV is the same as it is for the missile nozzle. There are several methods of improving the RV's resistance to the heat of atmospheric reentry.

With entry to the atmosphere at 90 - 100 km, as mentioned, there is what is called a "boundary layer transition", and there is a resulting increase in drag on the 7' high x 11.75" base diameter RV, and thus a decrease in velocity. The boundary layer of transition can be reduced by piping air through holes running close to the surface from near the top of the conical RV and exiting at its base.

Just below the hemispherical nose tip, evenly spread around the circumference are 85 small holes that duct the thin air of the upper atmosphere down through the RV's carbon skin, all the way to the base of the RV, where they are vented, making the RV more aerodynamic, and able to reach an altitude of 16 km/10 miles before atmospheric friction begins to heat the vehicle. [U.S.P. #4,185,558 (1980); classified 12 years] ["Countermeasures" (ca. 2000)]

The principle is similar to base bleed in artillery shells, where a burning charge at the shell's base makes the turbulent wake and eddy currents following behind the flat artillery shell base less of a drag on the shell. The wake is also a partial vacuum, tugging at the RV, and this is also reduced by the added air flow at the base.

The nose tip shape is rounded to produce a thick bow shock wave away from the RV nose tip, and reducing the absorbed heat.

Construction of the RV is made for heat conduction downwards away from nose tip. But first the pyrolytic graphite of the carbon-carbon composite nose tip preferentially conducts heat laterally (in the a-b flat plane) sacrificially absorbing and reducing/diluting it before conduction from the nose tip down to the lower surface body of the RV. In the c-direction – downwards through the thickness of the layer – the pyrolytic graphite is an excellent insulator, protecting the RV's inner warhead beneath its outer heat shield coating. [AD783 849 (1974)]

As well, due to Stefan-Boltzmann's Law, the hotter a material gets, the more and more it radiates the heat off as visible white light, and invisible UV light. The visible white hot incandescence of a descending RV is actually cooling off the RV, at the tremendous rate of $T^4$.

The first RV outer refractory coating was developed by defense contractor AVCO (U.S.P. #3,922,411; filed 1958 and issued 1975; 17 years classified), and consisted of sintered silica (SiO2/common sand) in a honeycomb layer of Inconel (mostly nickel) or molybdenum. The silica is formed by hot pressing at 1,200 ºC. at 2 - 3 kpsi for 1 - 2 hours. This was used in the Atlas and Titan I Mk 4 RV for the W-38 warhead, and the MM I Mk 5 and Mk 11 RV's. Beneath the refractory layer was a powder-forged beryllium W-38 warhead casing encased in a thin layer of stainless steel. [ORF50556 (1960)]

Then there was nose cone cooling. U.S.P. #3,682,100 (1972); filed in 1964, used a laminate of thin layers of alternating conducting metal, such as siliconized molybdenum with layers of a hydride, such as lithium hydride (LiH), to absorb the heat, and dissociate into hydrogen gas, which is ducted down to the base of the RV. The patent describes how LiH (and other compounds) is an excellent heat absorber for RV's. LiH has a high boiling point of 1,370 ºC, a high Specific Heat of 0.42 cal/cm3, and a high heat of dissociation of 2.23 kcal per gram of LiH. For a total heat absorbtion capacity (from heating to vaporization to dissociation and ionization) of 56 kcal/g.

It also gives the equation for the gas temperature near the surface of the nose cone by the following approximate formula:

$$T - T_0 = (v/100)^2$$

Where T is the temperature of the gas near the nose cone in ºC; $T_0$ is initial atmospheric temperature, and v = vehicle velocity in mph. As an example, with v = 10,000 mph = 16,000 km/hr, then T = 10,000 ºC.

The patent also suggested mixing graphite powder in with the LiH. Graphite sublimes at 3,500 ºC, and the heat of sublimation is 12 kcal/g. Graphite was the eventual premium choice of RV heat resistance from at least the first MIRV, on the Minuteman III in 1970 and beyond, with other SLBM's.

Then there was developed in 1961 a low radar cross section RV coated with 0.75" thick Micarta (melamine-fiberglass) over a fiberglass superstructure [U.S.P. #3,997,899 (1976), Chrysler Corp.; classified 16 years].

Silica-phenolic (called Refrasil), a charring ablator, is a composite developed during the early 1960's using high-purity silica fibers in a woven cloth, impregnated with phenolic resin. It has a density of 1.63 g/cm3, a specific heat of 1.2 kJ/kg-K, a thermal conductivity of 0.5 J/m-s-K, with an ablation temperature of 2,400 ºC, and resulted in an average depth of ablation on re-entry of less than 3 mm.

With a low density nylon-phenolic, a 2 cm thick shell of this material on a sphere with outer radius of 10 cm would have a mass of only 1.2 kg. Only 9 mm would be ablated, which though greater, the temperature profile shows that at 185 seconds, the inner surface of the heat shield would only increase by about 20 ºC.

\*　\*　\*　\*

" 'I have come to throw fire on the earth.
I wish that it had already started !' "

There are primarily five factors which affect the target accuracy of a ballistic trajectory (unpowered free-falling) Reentry Vehicle (RV):

1) Gravity,

2) the Earth's rotation,

3) Atmospheric affects (air pressure and density, water vapor content/humidity changes, winds/climate/clouds, resistance differences with height of RV at different altitudes. temperature differences),

4) Dynamics of the RV body, and

5) Asymmetries of the RV body, especially as its surface ablates (burns off); RV orientation and center of gravity changes. [AD731 662, p. 75 (1971)] [Cummings (1967)]

These effects are on top of the basic physics of the ballistic trajectory of the RV:

1) Velocity,

2) Time,

3) Flight path angle,

4) Altitude, and

5) Range.

There are two atmospheric warhead reentry vehicles meant to survive the tremendous air friction heat of atmospheric reentry to deliver the warhead safely and as accurately as possible to the intended target.

The first was the blunt-shaped, heat sink reentry vehicle (RV), and the second was the later-developed and much more accurate ablative, conical RV. Blunt-nosed, heat sink RV's were the simpler device, and were thus developed before the conically-shaped, ablative-type RV's.

H. (Harvey) Julian Allen and Dr. Alfred J. Eggers issued their first classified report on blunt body, heat sink reentry vehicles in 1951. Julian Allen is best known for his "blunt body" theory of aerodynamics, which counter-intuitively suggested the design on the grounds that a blunt body is slowed down by collision with air molecules. The blunt-nosed RV's had a low ballistic coefficient ($\beta$/beta; weight to drag ratio) of around 500 lbs/sq.ft., compared to as high as 3,000 for the conical-shaped, ablative RV's.

Atmospheric reentry also produced a hypersonic shock wave in front of the blunt RV, and actually shielded the vehicle from more heat than the later ablative conical RV's [Day 2003)] [U.S.P. #2,937,597 (1960)].

But the blunt-nosed, heat sink RV's had a lot of dead weight – hundreds of pounds – to absorb the heat of atmospheric reentry, compared to the much lighter ablative RV's, which had an effect on the range of the later, smaller solid propellant ICBM's/SLBM's.

The ablative was clearly superior for warheads. Its high beta made it faster, and thus much more accurate, as well as better prepared to avoid interception. This was also helped by the lower radar observability – a factor noted as early as 1960 – of the conical, ablative RV's, compared to the bigger blunt-nosed, heat sink RV's [AD-A241 725, 4-1 to 4-4 (1991)].

But the heat sink RV came first, as the needle-nosed high beta RV's burned up in wind tunnel tests. And the blunt body, heat sink RV was necessary for returning manned space vehicles for the civilian NASA space program, running almost in parallel with the military program.

Large weight of high heat absorption capacity, highly conductive materials used. Conduction was about equal to the heat convected away. The heat convected was also much greater than the heat re-radiated off. The bow-shock wave created ahead of the blunt body also reduced the frictional heating of atmospheric reentry, deflecting it, as well as slowing the RV's descent velocity, further reducing its heating.

The efficiency of copper is a heat capacity/specific heat of 0.092 cal/g.°C (270 BTU/lb.), and for beryllium it is the much better 0.436 cal/g.°C (1,500 BTU/lb.) ["CRC", D-141 (1972)], the second highest specific heat after lithium metal. And beryllium has the second highest heat of fusion of all. And the specific heat goes up fairly significantly with temperature (the values given are for 25 °C).

The blunt heat sink RV was made of an alloy of beryllium-copper and weighed several hundred pounds. The back surface of the frontal blunt heat sink RV was insulated from the warhead. ICBM Re-entry: peak heating rate of 2,000 BTU/ft2-sec for 0.1 seconds. So the heat sink was heated to several hundred degrees, which was less a problem with nuclear weapons, which could be insulated, than with manned vehicles. The blunt heat sink RV had a ballistic coefficient, $\beta$, of approximately 100 lb./ft2, (compared to 2,000 lb./ft2 for modern conical RV's) so that peak heating occurred near 50 km, and extended over a much longer time so that radiative heat transfer was adequate to dissipate the slower rate of heat generation; peak deceleration was on the order of 10g's. [AD-A229 778 (1990)]

The first heat sink design was the GE Mk 2 which had a low beta, beryllium heat-absorbing shield. Its front leading edge was a flattened cone, but it was not ideal, trailing a stream of ionized gas that showed up dramatically on radar, not a good thing for a warhead. Refrasil was perhaps used as the insulator between the Cu-Be heat shield and the warhead.

Refrasil, a charring ablator, had recently come into use as an external ablative RV spray coating. It consisted of high-purity silica (silicon dioxide) fibers woven into cloth, and impregnated with silicone or phenolic resin, and hollow micro-spheres. [U.S.P. #4,112,179 (1978)] ["Countermeasures" (ca. 2000)]. Refrasil had an ablation temperature around 2,400 °C, and good heat insulation properties. A sample reentry ablated only 3 mm of material.

Better insulating properties was with a low-density nylon-phenolic, which was lighter than a Refrasil layer, and with a 2 cm thick layer only allowed an increase of the internal temperature of 20 °C ["Countermeasures" (ca. 2000)].

The heat sink RV was used for the Thor IRBM (1959), Jupiter, Atlas, and a beryllium version in the Polaris A-1 and A-2 (1960 and 1962), and the Minuteman I (1962). The Polaris A-1 was an integrated warhead/RV design, using a hemispherical nose, a main cylindrical beryllium (Be) casing, with a flared section at the bottom for stability during descent, and two small rocket motors which spun the RV for stability and symmetrical erosion during reentry. The total weight of RV and warhead was about 700 lbs.

Refrasil was perhaps used as the exterior above the Be, and nylon-phenolic between the Be and the warhead.

For comparison, the first U.S. manned flight of the Mercury spacecraft in 1961 was a blunt-shaped reentry capsule with a 10 mil thick titanium skin, over an ablative fiberglass-phenolic heat shield. The astronaut-containing, cone-shaped Command Module for the Apollo 8, December 1968 launch of one of the series of NASA moon exploration vehicles also used a combination blunt-shaped/ablative reentry vehicle system.

It consisted of a honeycomb of brazed stainless steel, 0.7" thick, and filled with phenolic epoxy resin as an ablative material. The inner backing structure consisted of aluminum honeycomb bonded between layers of sheet aluminum 0.25 to 1.25" thick. Its prime contractor was Rockwell. [Yenne, p. 15, 19-20 (1985)]

But ablation systems were clearly the wave of the future, if they could be made to work. Unlike the ablation of the secondary in a staged nuclear weapon, the aim was the opposite. High heat resistance, high heat emissivity, and high heat conductance. The nose tip was the key. It was hemispherical in shape, sitting on top of a conical RV body. The hemispherical nose tip produced a hypersonic shock wave that stands off slightly in front of the vehicle. At still lower altitudes there was produced a (protective) "boundary [air] layer" encompassing the whole vehicle.

The nose tip produced what was referred to as the stagnation point, the region where all the kinetic energy of the air molecules, was converted into heat, so that the maximum temperature is attained here.

Charring-ablation systems work by heat dispersion by: re-radiation at the surface, decomposition of the plastic into gases and a solid residue referred to as char with the absorption of heat by the gaseous decomposition products, storage of heat in the remaining carbonaceous solid material. The char is composed of carbonaceous material containing reinforcing inorganic fiber to provide the strength required to withstand the mechanical forces present. [Loh, p. 128 (1968)]

Charring-ablations systems like phenolic-nylon (50% phenolic resin and 50% nylon fiber, by weight). Ablative material is injected into a fiber glass honeycomb matrix, bonded to stainless steel honeycomb substructure, and burns off in thin layers, taking the heat away with it. [Day (2003)]

14,000 BTU/lb. is absorbed in subliming at 1 atmosphere, which occurs during initial entry phase. During later phase of entry trajectory, surface erosion is due to oxidation.

RV ballistic coefficient is a parameter that designers try to maximize, since it determines the speed of RV atmospheric reentry. The faster the RV descends, the less it is affected by wind, rain, and other atmospheric effects, and the more accurate the RV is.

Ballistic coefficient is directly related to RV mass, and inversely related to the RV cross-sectional area (base area) multiplied by the drag coefficient (typically between 0.05 and 0.1). 2,000 lbs. per square foot is typical of modern conical MIRV RV's.

The speed of RV's, and thus their accuracy, improved from Mach 3 to Mach 10 (1 km/s) by 1971, with the W-68 Poseidon SLBM.

The first Soviet ICBM RV used the ablative method, with their usual simplicity, a large 15 MT H-Bomb with laminated plywood as a carbonizable ablative RV shell. [Stine, p. 184 (1991)]

For both weight and neutron hardness, in 1962, around the time of Minuteman I, and Polaris A-1 and Polaris A-2, they came up with plastic cases for the weapon. [RDD-4, II-Q-16 (1998)]. The design of RV and bomb had come to be seen as a single integrated design. The RV was beryllium, the casing with metal-ribbed plastic. Beryllium is a neutron reflector and moderator, as well as plastic too.

(U.S.P. #4,147,822 (1979) McDonnell Douglas) discloses a conical RV with a hemispherical nose, made of a framework of beryllium, with cut-outs, covered with a weave of fiber impregnated with a thermosetting resin. The graphite yarn (1.77 g/cc, with 1,500 filaments per ply had a tensile strength of 450 ksi, prepared from the precursor PAN, polyacrylonitrile, which is superior (because of its high carbon content, low cost, available in numerous filament count yarns and tows, and ease of conversion to high strength carbon fibers) than using rayon, which was used for 20 years) [AD-A116 828 (1982)] [AD-A033 540 (1976)], is wound on in a continuous strand by rotating the nose cone along its axis of rotation. Probably the RV for the small Poseidon warhead.

Then there was a thin zirconium dioxide coating over the C-C composite that acted as a superior incandescent radiator.

## Anti-ABM RV Measures

In the late 1960's to the 1970's nuclear warhead Anti-Ballistic Missile (ABM) missile systems were introduced by the Soviet Union (and then, in response, the U.S.; but the U.S. retired its ABM system in 1975, as well as having signed a reduced ABM deployment treaty in 1972 with the U.S.S.R.).

To defeat ABM systems, the U.S. introduced in the 1970 MM III ICBM and 1971 Poseidon SLBM reduced radar cross section RV's, higher reentry speed RV's, neutron and x-ray protection measures, and PENAIDS (Penetration Aids): decoys, chaff, dipoles, reflectors, or balloon-coated RV's.

For neutron/x-ray protection, in the RV, below the RV's outer skin, and surrounding the centrally-positioned warhead, was introduced a light-weight surrounding shell of Li7H (lithium7 hydride) to temporally stretch the length of the heat of the ABM's nuclear detonation bomb-thermal x-ray pulse, and attenuate its 14 MeV neutron flux, protecting the warhead from damage at a closer distance from the ABM's thermonuclear blast than a regular, unprotected RV.

The decoy radar cross section enhancer (U.S. Patent #4,700,190 (1987), declassified from 1979) used two dipoles antennas at right angles (orthogonal) to each other. Decoys – with a reduced size compared to real RV's – have a nose, ballast for weight, and RF enhancer to increase their radar cross section. Dipoles or reflectors are carried in a PenAids canister. Reflectors are conductive wires, and along with dipoles, have a specific classified length and diameter.

Chaff dispensers dispense separate, multiple chaff clouds – a "puff" of wires of a classified material and weight, and wire length and diameter. The cloud geometry (length and width) is also classified, as well as chaff ejection velocity, time of puff, and deployment angle.

There is also a BBU (Broad Band Unit) radar cross section PenAid, with a classified cloud geometry (length and width) and composition.

This is what they say regarding chaff for heavy bombers, such as the B-47 and B-52:

> "Chaff, as well as jamming, may be used to cover relatively narrow portions of the radio frequency spectrum or, less efficiently, to provide clutter over large frequency bands… Wide-band chaff consists of a number of lengths of chaff dipoles with or without rope, packaged together."

> "Only 7 lengths of dipole are required to cover the [enemy radar] frequency band from [2.5 – 10.5 GHz]…"

> "One length of U.S. standard foil dipoles is 36 mils wide by 0.45 mils thick, cut at S-band to yield [a length of 656 meters and ] 60 square meters of echoing area, [and] weighs about 2 ounces."

> "[E]fforts have been made to further reduce the bulk and [80% of the] weight of chaff by using narrower widths and by using aluminized glass filament dipoles in place of the foil dipoles." [AD354 894 (1959)]

**The Original Heat Sink RV's**

> "Jesus said to them, ' I watched
> Satan fall from heaven like lightning."
>
> -- Luke 10:18

Thor IRBM (1958)

GE Mk 2 RV weighed 3,900 lbs., was the same as Atlas RV; was massive conical shield of solid machined beryllium-copper, 0.625" – 0.75" thick and 6 ft. in diameter.  Behind the shield was a stainless steel compartment for a 1 MT bomb.  The shield alone weighed more than 1,000 lbs. of beryllium.  [Stine, p. 198 (1991)]

The Mk 4 blunt heat-sink RV weighed 950 lbs.  ["Report by Commander on the 1962 Pacific Nuclear Tests (Operation Dominic)", p. L-C-7-3 (1964)]

**Ablative RV's**

Two liquids, phenol and 40% (in water) formaldehyde gas combine via polymerization to form a hard thermosetting plastic, which forms an ablative matrix when reinforced with, e.g., a 2-dimensional nylon fiber weave.  This was the first polymeric composition for ablative RV's.

Jupiter IRBM (August 1958)

Ablative hemisphere-cone RV
ablative nylon-reinforced phenolic resin matrix
Dimensions:  65" base diameter x 9' high; 12.5" radius hemispherical tip
Weight:  1,017 lbs.
RV base shallow-dished (convex outward) bulkhead, essential for aerodynamic stability.
Greatest ablation was at nose tip, and was less than 0.375 inches.
Impact speed of RV:   Mach 0.45
Contractor:  Goodyear Aircraft Corporation  [Grimwood (1962)]

Atlas D/E/F ICBM (1959)

In the Atlas E/F an ablation-type Mk 3 RV, replacing the heat sink RV originally in the Atlas D.

Mk 3 RV ablative nylon-reinforced phenolic resin; nose cap and flare skirt; RV weight: 3,900 lbs., including its Mk 49 warhead.  [Goldberg, part 1, p. 873 (1981)]  Mark 49 warhead lengthened by 2.1", to 57.9" long, with an extra 92 lbs. of ablative material:  an increase from 1,640 to 1,732 lbs.  [RS 3434/30 (1968)]
Atlas D RV atmospheric reentry speed:  16,000 mph (Mach 22)
Contractor:  GE (General Electric) Corporation

Titan I ICBM (1960)

Mk 4 RV:  4' base diameter x 10.79' (129.49") height
Weight:  2,400 lbs.  [Stumpf, p. 31, 36 (2000)]
Contractor:  AVCO Corporation


Polaris A-1 (1960) and A-2 (1962) SLBMs

Mk 1 RV:  Hemisphere-cylinder-flare type RV; combination heat sink/ablative RV; used "G-1" beryllium hemispherical nose-cap, with cylindrical beryllium body, and a beryllium skirt welded on the bottom for aerodynamic stabilization.  It was an integral RV/warhead combination, the beryllium RV weighing around 250 lbs., with the warhead on the order of 600 lbs.

Contractor:  Lockheed
[Spinardi, p. 55 (1994)]  [Be2536 (1961)]

RV + W-47Y2:  717 lbs. and 18" dia. x 46.6" long with a 19.9" flared base; pyrolytic graphite, hemispherical nose tip, and insulation between the RV and the beryllium-cased warhead, to keep the temperature of the warhead to not more than 300 ºF.  [RS 3434/39 (1968)]


Minuteman I ICBM (1962)

Mk 5 (MM IA) or Mk 11 (MM IB) RV
Mk 5:  hemisphere-cone-cylinder-flare 78" high, cylindrical portion 19.5" dia., and flare 32" in
        diameter
Contractor:  Avco Corporation


Pershing IA (1962)  IRBM

Mk 4 RV with W-50 warhead, ablative sphere-cone RV; 950 lbs.
RV dimensions:  30" base diameter x 67" high; half-angle:  13º

Fiberglass conical superstructure coated with maximum of 0.75" thick Micarta (58% melamine resin/42% fiberglass) [Rolsma (1961) patent]  I guess you didn't know those thin, plastic counter-tops using Micarta also had a national security use!


Titan II ICBM (April 1963 – May 1987)

GE Mk 6 sphere-cone type RV:  Hemisphere-Tipped, Conical Body; warhead placed heavy end highest, so that RV has a high center of gravity for stability.  90" base dia. x 122.3" height (22.35" radius of sphere-cone nose); weight of RV alone was 1,375 lbs. or 1,800 lbs.; 10 ft. diameter domes made out of 2014-0 aluminum alloy.  [AD748416 (1972)]

Nose tip 2.3" thick maximum, ablative nylon-phenolic (Nylon-66 cloth impregnated with phenolic resin) (a charring ablator) 0.25" thick RV heat shield bonded onto aluminum honeycomb substrate.

Guidance: IMU (Inertial Measurement Unit), and heater, and MGC (Missile Guidance Computer). Bi-weekly tests included Dynamic Response Test, Target Select and Verify, Calibrate Test, Gyro Drift Test and Memory Hold Test, AAS (Azimuth Alignment Set) and IMU Alignment and Acquisition Check. [AD042 398 (1963)]

["Life" magazine (November 1987)]  [Stumpf (2000)]  [National Atomic Museum web page, circa 2002] [U.S.P. #4,340,197, Campbell, filed Feb. 1966, declassified 1982] [Hansen, VII-352 (1994)]


Polaris A-3 SLBM (1964)

3 W-58 MRV "cluster warheads"; Mk 2 ablative RV; beryllium nose tip; nylon fiber-phenolic ablative heat shield bonded to aluminum superstructure carrying the W-58 warhead. [Spinardi, p. 67 (1994)] With a BeO antenna window. RV: 23.5" base flare diameter x 54" high; weight of RV + warhead: 300 lbs.; hardened RV MRV's ejected at 200,000 feet altitude.


Minuteman II (1966)

Throw-weight of 1,100 lbs. [Buchonnet (1976)] (RV + W-56) weighed 1,100 lbs.; W-56 weight 560 lbs.; RV weighed approx. 540 lbs. [WSEG Report No. 50 (1960)] Beta equals approximately 1,000 lb./ft2 [AD731 662 (1971)]


**MIRV and other Advanced Reentry Vehicles (1970 Onwards)**

> "[Two apostles] asked, "[Jesus,] do you want us
> to call down fire from heaven to burn them up?"
>
> -- Luke 9:54

MIRV stands for Multiple Independently-targeted Reentry Vehicle: multiple nuclear warheads on a single ICBM, capable of hitting several different targets. The original concept was conceived in 1962 by a physicist at the RAND Corporation, Richard Latter, who had previously set up the Theoretical Division of the UCRL in the early 1950's. In 1962 the Mk 12 MIRV was placed under development, for an IOC by 1970 on the Minuteman III ICBM.

In October 1963, the U.S. civilian space program – NASA – had independently already launched a rocket booster which dispensed multiple satellites into Earth orbit, which was similar in technology to the planned MIRV. [Kaplan, p. 360 – 363 (1983)] [U.S.P. #3,380,687 (1968)]

MIRVs were originally conceived for:


1) force multiplication – to hit many more targets with the same number of ICBMs, and

2) to overwhelm enemy ABM defenses, nullifying their usefulness.

It was first successfully tested in 1969. It was actively promoted by ex-LLNL scientist, conservative DDR&E Director John Foster, who reigned from 1965 – 1973, during which time MIRV warheads were first deployed on the MM III ICBM (1970), and the Poseidon SLBM (1971), and was made possible by the light weight and small size of the latest two-stage weapons, the 170 kt W-62 for the MM III, and the 40 kt W-68 for the Poseidon SLBM.

This allowed multiple RV's – now cone-shaped warheads – to be launched with one missile, and which would hit individually separated targets (with a relatively large, elliptically-shaped "footprint" of possible strikes) with much higher accuracy. The elliptically-shaped footprint has its major (long) axis matching the trajectory of the "bus". [AD-A090 151 (1980)] The accuracy came not just from the superior reentry aerodynamics of the conical RV's, but from the bus which dispensed the MIRVs.

The conical MIRVs had a high center of gravity, by design. For stability, the heavy end of an internal nuke is placed close to the pointy apex end of its RV.

The bus – Post-Boost Control Subsystem (PBCS) RV launch vehicle -- was essentially a liquid-fueled fourth stage at the top of the MM III ICBM, carrying and individually launching the three MIRVs as the bus & MIRVs alone approached the target area.

The MIRV technology was independently consistent with the civilian, NASA-developed, Tran-stage Titan III method of launching multiple satellites, [Leitenberg, p. 10-11 (1984)] which had been demonstrated successfully in October 1963.

But U.S. designers experienced serious problems in designing the MIRV bus, which dispensed the MIRV warheads one-by-one at separate targets. The problems in development included stopping the bus from oscillating with the recoil of dispensing each MIRV.

In order to evade Soviet ABM (Anti-Ballistic Missile) systems, there were several factors. These anti-ABM methods included increasing penetration speed, reducing radar cross section, using decoys and chaff, and altering the trajectory using maneuvering aerodynamics technology.

To utilize this last method were developed rolling mass designs, which incorporated an internal mass, which shifts, thus causing a shift in the center of gravity, which in turn changes the angle of attack by changing the trajectory. It was the most cost-effective MaRV design, saving both weight and fuel. The maneuver could be carried out several times, more efficiently at a lower RV altitude (at high altitude the atmosphere was too thin), and generating a highly unpredictable trajectory. [AD-A440 289 (2005)] [U.S.P. #4,577,812 (1986); classified 12 years)]

In addition, wind and atmospheric variations affected the accuracy of the Mk 12 MIRV on its descent to the target. Guidance miniaturization was also an issue for the bus, into which it would have to fit. [Rovner (2011)]

Reducing the base diameter of the MIRV cones, allows sharper pointed RV's (still with a rounded nose tip, though!) that descend faster and were much more accurate, due to more immunity from wind, rain, and other atmospheric distortions. Thus there was pressure on the design labs for more diameter reduction: smaller two-stage H-bombs had smaller yield, which could be increased by using more of the more expensive tritium and Oralloy, but allowed for more accurate high-beta sharper pointed RV's; higher speed get hotter, design thicker ablative RV's.

An energy deposition of 20 kJ/cm2 (4,500 cal/cm2) evaporates (ablates) a layer of carbon 3 mm thick. [Bethe et al. (1984)] The total reentry energy deposition we are dealing with is ~800 kJ/cm2.

Pyrolytic graphite has a high thermal conductivity in the lateral (a-b) plane, a low thermal conductivity in the c-plane (through the thickness of the layer), and is a heat-sink material. It has the highest strength of any known material at temperatures above 3,000° F, and it is about ten times stronger than tungsten at

5,000 °F.  It also has good ductility at higher temperatures, even though it is quite brittle at regular temperatures.  [AD368 640 (1965)]  It was used in MIRV nose tips to conduct heat away, and further down the MIRV body, diluting it, and allowing it to be radiated away very efficiently at a rate of $T^4$.  Thus doubling the temperature of the pyrolytic graphite results in radiation of 16 times the amount of energy.

The original design of the MIRV nose tip was a hemispherical tip on top of a conical lower portion.  Nose tip performance was directly related to accuracy and survivability.  Many of these problems are directly responsible with nose tip performance.  [AD-A066 217 (1978)]

The HIP (Hot Isostatic Press) subjects materials to high heat and even pressure.  Used for compacting RV pyrolytic carbon making it harder, smooth, and more heat resistant, especially RV nose tips, making for more accurate RV's.  The HIP is a dual-use item, subject to COCOM export controls, if it can be used for items more than 5" in diameter.

The pyrolytic graphite nose cone was prepared using a mandrel for shaping.  The mandrel is prepared by hand polishing of the surface with 600 grit paper, washing with volatile organic solvents, and then depositing an ultra-thin pre-coat (0.00025" to 0.0005") for ease of separation from the mandrel.

Graphite's strength is unequaled at high temperature, by any other known material, as well as having the highest melting point of any element.  It has low weight, good heat resistance and ablative properties.

The flow field around a decelerating conical RV can be divided into five regions:  the stagnation region, intermediate region, aft-body region, wake region, and the boundary layer.

The wake region is the turbulent, low pressure region that follows the base of the cone.  (Addressed by a series of small tubular holes extending from near the top of the RV to the base of the RV [U.S.P. #4,185,558 (1980)].)

The stagnation region exists at the front of the vehicle, and contains the high temperature and pressure shock wave.  The viscous boundary layer exists along the conical surface of the entire vehicle, except the base.  Like any boundary layer, a large gradient exists in temperature and velocity through it.  [Starkey (2003)]

Carbon-carbon composites have superior thermal and mechanical properties that make them ideal for temperatures above 2000 °C.  C-C composites (CCC) consist of a carbonaceous matrix (pyrolytic graphite) reinforced by carbon or graphite fibers having at least a two-dimensional fiber in resin weave system.  The carbon fibers make the mechanical strength of the composite so strong.

CCCs were discovered by a chance observation.  In 1958 a technician at Chance-Vought's Aircraft Astronautics Division (later, Ling-Temco-Vought, LTV) experimented with burning a fiber-reinforced phenolic matrix composite, inadvertently covering it during the pyrolyzing process.  This resulted in a porous composite, composed of carbon fiber in a porous carbon matrix, which had a high impact resistance and strength.  [AD-A528 970 (2005)]

High density carbon-carbon composite (CCC) is the basis of the modern MIRV RV coating.  It is manufactured by depositing pyrolytic carbon on a wound fibrous carbon filament base:  carbon has the highest melting and boiling point of all elements (actually it sublimes rather than melts).  And pyrolytic graphite has one of the highest heat capacities of all, and is a good conductor in the lateral/horizontal plane.  It's sublimation energy is 191 cal/g.  It has a low ablation rate and high strength at elevated temperatures.

CCC's are very strong and stiff;  the strength is uni-directional along the longitudinal axes of the flat fiber layer, so multiple layers are required for strength along the X-Y axis for 2-dimensional material, which is what is required.  The even better 3-dimensional CCC was first developed in 1965 by Sandia Labs and scaled up to missile reentry vehicle heatshield sizes, and used successfully.  During the mid 1970's the Air Force focused on the development of 3-D shape-stable CCC RV nose tips.

The tensile strength, already high, increases with temperature up to about 1,650 °C, and then decreases with further increase in temperature. The number of filaments per strand generally ranges from 2,000 to 12,000. The carbon-carbon composite was fine-weave (700 filaments per yarn), and processed with pitch at impregnation pressures of 10,000 psi to a density of 1.9 g/cm3. Finally, graphitization is performed at 2,700 – 3,000 °C. [AD-A033 540 (1976)]

Carbon-carbon composites were first developed in the U.S. starting in 1958, by Chance Vought Aircraft Inc., and a year or two later improved considerably with the advent of commercially available graphitized viscose rayon fabric, with the resulting all-carbon composite. The process for graphitizing woven and non-woven cellulose fabric (like rayon) had a patent filed in 1960, U.S.P. #3,107,152. [AD-A325 314 (1996)]

High-modulus (6 million psi) , high-strength carbon fibers are produced by the pyrolysis (vacuum heating) of relatively inexpensive, commercially available polymeric textile fibers. The two precursor fibers which were most frequently used were Rayon cellulose or PAN (polyacrylonitrile). During the oxidation and pyrolysis of the fibers, the carbon atoms rearrange into a near graphite-like structure. [AD742 765 (1971)] The density of pyrolytic graphite is 2.0 - 2.1 g/cm3. The carbon fibers are composites composed of graphite crystallite and organized carbon. The fibers are typically 92% or higher in carbon content, and from about 5 - 15 microns (µm) (0.20 – 0.59 mils) in diameter.

For fibers with a 100 million psi modulus, the ribbons are typically about 30 layer planes thick and 90 Angstroms wide. The fiber structure is described as an outer "onion skin" (circumferential orientation) with a spoke (radial) core. [AD773 168 (1973)] [AD-A043 064 (1977)]

Carbon fiber was prepared from 2-dimensional (cross-weave) woven polymer fiber, such as rayon (cellulose) and PAN (polyacrylonitrile) (with a diameter of 7 µm), by carbonization, graphitization, and densification. This formed the base (58% weight) for a carbonaceous filler (8%) and 34% by weight, high temperature phenolic resin binder.

Pyrolytic carbon is formed through the cracking or pyrolysis of gaseous hydrocarbons, such as methane:

$$CH_4 \rightarrow C + 2H_2$$

Pyrolytic Graphite (PG) is produced by vapor deposition techniques: the thermal decomposition of the hydrocarbon gas (4 - 14 % methane gas was used) on a heated substrate, such as carbon fibers (200 hrs. at 1,700 °C and 3 mm Hg pressure; with a methane gas flow rate of 10 cubic feet per hour). [AD600 907 (1961)] [AD824 697 (1967)] The orientation of the deposited PG is parallel to the carbon fiber surface layer plane. This vapor-deposited form of graphite has a very high purity, a high density (extremely low porosity), and a highly oriented crystalline structure. [AD401 115 (1963)] Because of the highly oriented crystalline structure, PG is an excellent conductor along its surface, and at the same time a good insulator across its thickness.

This is because PG is a highly oriented, highly anisotropic material without a binder phase, and having almost no porosity. The c-direction (thickness plane) strength of the PG (i.e., across the oriented crystallite layers that are in the a-b plane (flat plane) is both the lowest and generally the most critical in RV applications. One approach to improving the c-direction strength is the introduction of boron, which may pin the crystallite layers together. [AD850 572 (1969)]

Its reentry vehicle nose cone properties are improved as a boron-alloyed pyrolytic graphite, and is conveniently produced by adding the gas boron trichloride to the inlet flow of methane (or natural gas), at 3,750 °F, for on the order of 50 hrs. and at a pressure of 2 mm Hg. [AD824 697 (1967)] Another additive is silicon carbide (SiC). PG/SiC is produced in a gaseous codeposition process, the SiC depositing as needles perpendicular to the PG deposition surface, providing reinforcement in the c-direction, and vastly improved shear strength between the a-b layers of the pyrolytic graphite matrix. The amount of SiC is 15 - 20 weight % of the PG. [AD783 849 (1974)]

In solid propellant rockets, pyrolytic graphite washers are stacked so that their internal diameters form the nozzle throat. The PG erosion rate with a 6,000 °F propellant, with a motor pressure of 1,000 psi was about 0.4 mil/sec.

CC Composites are fiber-reinforced composites of carbon fibers and a matrix of pyrolytic graphite formed by graphitizing petroleum or coal tar pitch by heat treatment above 2,000 °C. CCC is stronger and stiffer than steel and lighter than aluminum.

Pyrolytic graphites are formed by vapor deposition techniques, resulting in a highly oriented, highly anisotropic material without a binder phase and having almost no porosity. The c-direction strength of the PG materials (i.e., across the oriented crystallite layers that are in the a-b plane) is both the lowest and generally the most critical in advanced reentry system applications. One approach to improving the c-direction strength is the introduction of boron, which may pin the crystallite layers together.

Monolithic graphites are made up of carbon crystallites in a carbon binder, generally a pitch-like material. [AD850 572 (1969)]

Pyrolytic carbon was developed as a more practical material to use for RV's than pure graphite, which can't be used because it's a powder that can't withstand mechanical stress, but would be an excellent (i.e, the best) RV coating if it could be.

There are three main forms of carbon: regular carbon (coal, charcoal, and coke), graphite (pencil leads and lock lubricant), and diamonds (jewellery and abrasives). The structure of graphite consists of covalently bonded tiny, flat sheets of carbon stacked in an …A-B-A-B-A-B… sequence, with van der Waal's force the only bond between sheets, which can slide back and forth on top of each other. It's this unique structure that gives it the "anisotropic" nature that gives it different heat conductivity depending on what is the direction of the stacked, aligned crystals. Pyrolytic graphite was developed to be as close in thermal properties to graphite as possible, but usable as a lasting structural material instead. [AD688 074 (1969)]

Energy absorbed by the RV include absorption by carbon's high specific heat, high heat of varporization, and conductivity, which distributes the nose tip heat throughout the RV surface.

It is produced by the thermal decomposition of the hydrocarbon gas on a heated substrate, such as carbon fibers. Its reentry vehicle nose cone properties are improved as a boron-alloyed pyrolytic graphite, and is conveniently produced by adding the gas boron trichloride to the inlet flow of methane (or natural gas). The boron alloy is 0.5% boron by weight. [AD824 697 (1967)] [AD783 849 (1974)] [AD688 074 (1969)]

In solid propellant rocket nozzles, pyrolytic graphite washers are stacked so that their internal diameters form the nozzle throat. [AD392 561 (1968)]

The ideal throat insert material for solid rocket motor application is a perfect insulation in the radial direction, has a melting temperature above the flame temperature, is chemically non-reactive, resists mechanical erosion, and is structurally isotropic. In the real world, such a throat insert material does not exist. CC Composites come closest. Due to the high conductivity of pyrolytic carbon, a surrounding layer of beryllium oxide or beryllium metal would improve its performance.

CC Composites are fiber-reinforced composites of carbon fibers and a matrix of pyrolytic graphite formed by graphitizing petroleum or coal tar pitch by heat treatment above 2,000 °C. CCC are stronger and stiffer than steel and lighter than aluminum.

Curing at 160 °C and 1,000 psi, the material was carbonized at 1,000 °C. under an inert gas flow. The overall processing time of slow heating and cooling rates encompassed 62 hours, followed by graphitization was performed at 2,600 °C for 8 hours. in an inert atmosphere. The phenolic char was partially converted to glassy carbon during the carbonization phase.

The densification phase was performed by impregnation with coal-tar pitch and carbonization in a Hot-Isostatic Press at 1 kbar and 650 ºC. Density 1.4 g/cm3

The performance of 0.55 mm was about 8,700 J/cm2. 33 kbar tensile strength of carbon fibers; density 1.8 g/cm3; m.p. 3,550 ºC [ADA216 540 (1989)]

There are three forms of carbon, all of a class of materials referred to as polycrystalline materials. Graphite is an allotropic crystalline form of carbon, referring to a hexagonal plane, stacked layered structure, with the stacked layer bonding much weaker than the hexagonal plane layer.

Pyrolytic graphite is manufactured by atom by atom deposition on the heated substrate, and has a high degree of crystalline orientation and produced high purity, high density and zero porosity, with an orientation parallel to the substrate surface, and a marked degree of anisotropy in thermal and mechanical properties.

The term "carbon substrate" is used to mean any bonded or compacted carbonaceous material containing at least 90% carbon, preferably graphite for its thermal stability, or carbon fiber, with a negligible erosion rate.

\*   \*   \*   \*

The following list shows the evolution of RV structure and RV materials as chronicled by the frequently classified U.S. Patents:

U.S.P. #3,026,806 (filed 1957, declassified 1962). "Ballistic Missile Nose Cone"; copper or MgO preferentially ablate off, absorbing heat, and being an envelope of cooling vapor.

U.S. P. #3,922,411 (filed 1958, declassified 1975; 17 years classified, AVCO). Coating of sintered silica ($SiO_2$) in a honeycomb layer of Inconel (mostly nickel) or molybdenum.

U.S.P. #3,095,162 (filed 1959, declassified 1963). First use of pyrolytic carbon: an excellent conductor pyrolytic carbon is the material deposited on a substrate by the thermal pyrolysis of a carbon-bearing hydrocarbon vapor.

U.S. P. #3,860,445 (filed 1959, declassified 1975). Carbide on graphite surface for increased RV heat resistance; metallic carbide of molybdenum, niobium, or zirconium.

U.S.P. #3,993,738 (high-strength graphite).

U.S.P. #3,682,100 (filed 1964, declassified 1972). Cooling of RV nose cone with hydrides. $ZrH_2$ + MgO. RV: $T - T_0 = (V/100)^2$; if V = 10,000 mph, $T_0$ = 0º C, then T = 10,000º C

U.S.P. #3,317,338 (filed 1964, declassified 1967). "Pyrolytic Graphite Coating Process"; coating of a substrate with pyrolytic carbon, by heating a base carbon substrate (91 - 98 wt.% carbon) from 1,900 - 2,400º C., and the heated substrate coated by passing over the substrate a lower aliphatic hydrocarbon gas, such as dilute methane/nitrogen mixture. The substrate consists of carbonized cured furan resin.

U.S.P. #3,410,746 (filed 1964; declass. 1968). "Grain-Oriented Pyrolytic Graphite Forms and Method of Making Same".

U.S.P. #3,369,920 (filed 1964; declass. 1968). "Process for Producing Coatings on Carbon and Graphite Filaments"; 40 patents reference this one (2000).

U.S.P. #3,412,062 (filed 1965; declassified 1968). "Production of Carbon fibers.."; carbon fibers having a high tensile strength are formed by carbonization at 1,000 °C. in an inert atmosphere, after heating to 200 °C. in an oxygen atmosphere while held under tension.

U.S.P. #3,629,049 (filed 1966; declassified 1971). "Shaped Pyrolytic Graphite Articles"; three-dimensional objects woven.

U.S.P. #3,657,061 (filed 1966, declassified 1972). A laminated reinforced woven fabric formed from graphite fibers, produced by weaving fibers so that they run in three dimensions orthogonal to each other.

U.S.P. #3,547,676 (filed 1966, declassified 1970, AEC). Synthesis of dense, isotropic pyrolytic carbon nose cone material.

U.S.P. #3,778,300 (filed 1966, declassified 1973; Los Alamos, AEC). "Method of Forming Impermeable Carbide Coats on Graphite"; applying a mixture of sodium nitrate and a refractory carbide (such as hafnium, niobium, vanadium, etc.), heating at 1,350 °C. in an inert atmosphere, then continuing heating in a vacuum at 2,300 °C.

U.S.P. #3,578,030 (filed 1967; declass. 1971). "Ablative and Insulative Structures"; a structure that is a body of revolution is formed by laying up side-by-side panels of high temperature resistant materials impregnated with thermally curable resin.

U.S.P. #3,607,541 (filed 1967; declass. 1971). "Process for Producing Pyrolytic Carbon Impregnated Thermal Insulating Carbonized Felt"; a low density carbon felt of carbonized viscose rayon is subject to pyrolysis with methane at a reduced pressure and 1,800 °C.

U.S.P. #3,639,158 (filed 1967;declass. 1972). "Structural Carbonacerous Materials having Improved Surface Erosion Characteristics"; graphite fiber impregnation with high boiling point resin.

U.S.P. #3,644,222 (filed 1967; declass. 1972). "Ablative Epoxy Resin Composition"; cured epoxy resin.

U.S.P. #3,635,675 (filed 1968; declass. 1972). "Preparation of Graphite Yarns".

U.S.P. #3,644,135 (filed 1968; declass. 1972; TRW). "In-Situ Carbiding of Pyroyzed Composites"; impregnation of high temperature carbonaceouse filament bundles with refractory metal followed by carburizing.

U.S.P. #3,635,675 (filed 1968;declass. 1972). "Preparation of Graphite Yarns"; a method of making high tensile strength graphite fibers, by heating a synthetic polymer yarn to 2,000 °C for five minutes, with a pre-treatment of 350 °C in an oxidizing atmosphere.

U.S.P. #3,853,586 (filed 1968, declassified 1974, ARCO). "Tapered Carbon/Pyrolytic Graphite Composite Material"; rocket nozzles are improved by the tapering of a liner of high tensile strength, and low CTE (coefficient of thermal expansion) pyrolytic graphite, with a thicker center section, tapering towards its ends

U.S.P. #3,862,334 (filed 1969, declassified 1975). "Method of Manufacturing Carbon Fibers"; pretreatment of a polymer fiber by heating in an inert atmosphere at 200 - 300 °C. before heating to 200 °C in an oxygen atmosphere, followed by pyrolysis at 1,000 °C in a nitrogen atmosphere.

U.S.P. #3,885,077 (filed 1969, declassified 1975, McDonnell Douglas). Production of uniformly expanded pyrolytic graphite.

U.S.P. #3,724,386 (filed 1970; declass. 1973; Air Force). "Ablative Nose Tips and Methods for their Manufacture"'; a nose tip consisting of a unidirectional rod of carbon fiber and CVDed phenol-formaldehyde resin.

U.S.P. #3,798,161 (filed 1970, declassified 1974). "Composition for Preparing Graphite Bodies"; forming into the final body shape petroleum coke with isotruxene and acenapthylene as the binder, then firing to 2,600 ºC to completely graphitize to a porous form suitable for refractory metal carbide coating.

U.S.P. #3,875,106 (filed 1970, declassified 1975, McDonnell Douglas). Silica fibers with a boron oxide-epoxy mixture to char/ablate uniformly.

U.S.P. #3,596604 (1971, Air Force). A pyrolytic graphite nose cone tip, made of stack of pyrolytic graphite blocks; pyrolytic graphite is carbon deposited from a hydrocarbon vapor over the temperature range of about 2,000 ºC; it is a specific high temperature form of pyrolytic carbon.

U.S.P. #3,819,461 (filed 1971, declassified 1974). "Unidirectional, High Modulus Knitted Fabric"; graphite fiber cloth; 29 patents reference it (in year 1999).

U.S.P. #3,895,084 (filed 1972, declassified 1975, Duocummun Inc.). "Fiber Reinforced Composite Product"; 33 patents reference it (in 2000); see related U.S.P. #3,991,248.

U.S.P. #3,960,626 (filed 1973, declassified 1976, Martin Marietta). "Method of Making High Performance Ablative Tape"; used in the Trident I RV, as TWCP (Tape-Wrapped Carbon-Phenolic); uses non-woven, narrow (inch-wide), high temperature resistant tape, made from a longitudinal array of parallel carrier carbon fibers and a large number of short fibers that are not woven into the carrier fibers, but reside in a dense manner upon the carrier fibers, sticking out in an orthogonal orientation; the short, brush bristle-like short fibers are glued to the carrier fibers with, e.g., phenolic resin; the tape is wound around the RV substructure and a plastic polymer, such as phenolic resin is applied and cured.

U.S.P. #3,991,248 (filed 1974, granted 1976, Ducommun). "Fiber Reinforced Composite Product"; 64 patents reference it (1999); preparation of carbon-carbon composite (carbon fiber infiltrated with pyrolytic graphite) with accurate control of density, shape, fiber volume and orientation; a stack of cut panels of graphite cloth is compressed between two metal plates, removed from the compression fixture and infiltrated with pyrolytic carbon.

U.S.P. #4,218,276 (filed 1974, declassified 1980, Avco). "Method for Making 3-D Structures"; graphite cloth is stacked on the lower metal plate of a jig containing upright needles; the needle holes are eventually filled with filamentary graphite fibers, and the cloth impregnated with phenolic resin.

U.S.P. #3,914,395, and U.S.P. #3,993,738, in 1975 and 1976, respectively, disclose a method for production of high strength graphite fibers by pyrolyzing rayon or other plastic fibers at 700 ºC. in an inert atmosphere to carbonize them, followed by resin impregnation, then graphitizing at 2,750 ºC. in an inert atmosphere.

U.S.P. #4,100,322 in 1978, disclosed production of low weight carbon fiber/phenolic resin composites, prepared by impregnating graphite fiber cloth with phenolic resin, curing the resin, pyrolyzing the composite at 1,600 ºF in an inert atmosphere, and then re-impregnating the porous carbonaceous matrix with additional phenolic resin.

U.S.P. #4,152,381 in 1979, disclosed how to produce metallated graphite filament-wound structures, by coating a continuous multi-filament carbon cloth, with a metal carbide, impregnating the carbide-coated cloth with a polymerizable carbon precursor resin, such as phenolic resin -- winding the resulting filament around a mandrel, partially allowing the precursor to cure at room temperature, then heat and pressure to fully cure it, and finally heating and pressurizing the structure at at least 1,000 psi to graphitize the polymer resin.

U.S.P. #5,004,590 (filed 1983, declassified 1991, Hercules). Carbon fibers from filamentary polyacrylonitrile (PAN) with tensile strength of 600 - 900 kpsi.

U.S.P. #5,368,076 (filed 1985, declassified 1994). "Process and Apparatus for Manufacturing Rocket Exit Cones and the Like".

U.S.P. #5,641,366 (filed 1988, declassified 1997, Loral Vought). "Method of Forming Fiber-Reinforced Composite".

U.S.P #5,686,027 (filed 1989, declassified 1997, Aerojet-General). "Process for Forming Carbon-Carbon Composite"; cross-linking polyphenylene resin produced unusually thick char yield on RV.

U.S.P. #4,949,920 (1990, Navy). "Ablative Cooling of Aerodynamically Heated Radome"; an RV nose tip cooled by transpiration ablative cooling of thallium metal, which is inside the RV, melted (heat of fusion) and vaporization (heat of vap.), and seeps through the porous nose tip It mixes with the air in the flow stream to form a protective boundary layer extending downstream the body of the RV.

U.S.P. #5,108,830 (filed 1991, Navy). "Shape-Stable Reentry Body Nose Tip"; carbon matrix nose tip having a metallic filament infused central core; manufacturing process involving 3-D weave of fibers and metallic tungsten filaments in the Z-axis only (vertical), vacuum impregnation with pitch, 800 ºC carbonation, followed by graphitization, then vacuum impregnation with pitch again, followed by pressure impregnation, then 650 ºC pressure carbonization, followed by graphitization, and final machining to shape; for Trident II.

U.S.P. #5,413,859 (filed 1992, declassified 1995, Lockheed). "Sublimitable Carbon-Carbon Structure for Nose Tip for Re-Entry Space Vehicle"; hemispherical carbon-carbon nose tip covered by weave of refractory metal fibers, such as tungsten or titanium or niobium; carbon-carbon inner coating of Teflon; for Trident II.

U.S.P #5,336,520 (1994). Commercial porous graphite is HIP pressed in a refractory metal container to produced a solid graphite piece with a density of at least 2.10 g/cm3; the HIP pressing is in an argon atmosphere, at 2,200 ºC, and 30 ksi for up to eight hours.

"A typical re-entry body [REB] has the shape of the frustum of a cone. It Consists of an aeroshell and whatever is contained inside the aeroshell. If the re-entry body is a nuclear weapon, the aeroshell contains a nuclear warhead. The aeroshells for most nuclear weapons consist of an aluminum shell that is ~.060-inch thick over which is placed a much thicker shell of ablative material. A layer of adhesive connects the two shells. The aeroshells that contain the W62, W76, W78, W87 and W88 all have this basic structure." [SAND94-0489]

The aluminum aeroshell has internal circular doughnut-shaped "ribs" parallel to the REB base to reinforce aeroshell rigidity and structural integrity.

Minuteman III (1970)

Mk 12 RV, upgraded in 1979 to Mk 12A RV, with a much higher yield warhead, and added an improved guidance system which gave the ICBM a "hard target" kill capability [AD-A223 379 (1986)]

(RV + W-62) weighed approximately 200 kg. [Buchonnet (1976)] A high beta RV. Spins, like all RV's Height: 59.9" (1.8 m); Base Diameter: 17.25"; Half-angle: 8.2°

Mk 12 mid-section of the aeroshell containing the W-62, frustum of a cone (a truncated conical segment: a cone with the top loped off, so that the cone is flat-topped rather than "pointy"), 22" high, and 17.27" in diameter at the base, and 8" in diameter at the top.

The aeroshell structure was 0.5 inches thick (0.41 inches carbon fiber-reinforced phenolic, glued with 0.040" Epon 934 (a silicone adhesive) to an aeroshell substructure of 0.060" thick (7049-T7352 aluminum), making the warhead cavity dimensions a maximum of 22" high, and a minimum of (8 - 1 = ) 7" inches in diameter. There were four small radar windows near the aft end (i.e., near the base) of the Mk 12 RV.

An upgrade in 1979 replaced the Mk 12 RV with the Mk 12A in 300 MM III ICBM's. The Mk 12A was essentially the same as the Mk 12, but with a perfectly conical outline at the top fore-end, and no radar windows, but an increased ballistic coefficient of 2,000 lbs./sq.ft. [Bunn (1984)] The RV's consist of a light-weight aluminum frame, its surface mostly covered by a carbon-phenolic heat shield, except for the hemispherical RV nose tip, which absorbs (and conducts down the RV surface) most of the reentry heat.

|  | Nose Tip | RV Outer Heat Shield |
| --- | --- | --- |
| Mk 12 RV: | Teflon (PTFE) over carbon-phenolic 5055A | Carbon-phenolic 5055A |
| Mk 12A RV: | Carbon-carbon | Carbon-phenolic 5055A |

[Hove, VI-173 (1979)]

The Mk 12A hemispherical nose tip is a bonded assembly comprised of three components, all made by AVCO:

> a) the shell, nose, forward section, is made of machined 7075-T73 aluminum, with attaching threads for the nose tip assembly;
>
> b) a nose shield, a 2-dimensional carbon/phenolic fairing between the nose tip billet and the RV; and
>
> c) the actual nose tip itself, a 3-dimensional, fine-weave pierced fabric (FWPF) carbon-carbon machined billet. The raw carbon-carbon nose tip was densified – or high pressure impregnated/carbonized (HiPIC) – by GE.

FWPF carbon-carbon billet fabrication: 7" x 7" squares of graphite fibre fabric are stacked, pierced by steel rods, and then compacted until a specified height of stack is reached. The compacted stack is removed from the piercing fixture, and the steel rods removed one-by-one, and replaced by graphite fiber rods. 1,000-filament graphite yarn is used to fabricate the woven fabric, and 3,000-filament graphite yarn is used for the rods. The final density of the carbon-carbon billet is 1.97 g/cm3, and it has an ultimate tensile strength of 27,000 psi., and a compressive yield strength of 17,000 psi, and a thermal conductivity in the x-direction (downwards along the RV skin) at 1,500 °F of 600 BTU in/hr. ft2 °F.  [AD-A098 593 (1980)]

When all the steel rods have been replaced, the preform is immersed for a specified time in coal-tar pitch at 500 °F, and then drained and heat-treated at 1,200 °F in an inert atmosphere.  It is then graphitized at a temperature of 4,700 °F.  After cooling, the rigidized preform is machined into 4 nose tip preforms.  The preforms are then densified, which consists of high pressure impregnation-carbonization (HiPIC) and a subsequent carbonization.  Precise control of temperature and pressure during HiPIC are critical.

The 2-D carbon-phenolic nose collar/thermal shield is made from filled, phenolic resin-impregnated carbon fabric.  The 3 components are adhesive-bonded together with clamping.  [AD-A098 595 (1980)] The structure of the MM3 aeroshell is similar to that for the W76 (Trident I), W78 (MM3 upgrade), W87 (MX/Peacekeeper), and W88 (Trident II).  [SAND94-0489] [SAND94-0335]  The patent U.S.P. #4,613,522 (1986), declassified after 4 years, by AVCO.  AVCO contractor for ABRV (later called the Mk 21) for  the MX/Peacekeeper missile.  Oxidation Resistant Carbon-Carbon Composites.

Poseidon SLBM MIRV (1971)

Mk 3 RV:  12" base dia. x 74.4" height; Nose tip:  hemispherical, 1" in dia.; ATJS graphite by Union Carbide; half-angle:  4.25°; Weight:  160 lbs. (RV+ W-68); RV was a heat sink/shield design made of beryllium, with an ablative graphite nose tip.  With a silica antenna window.
Reentry velocity:  25,200 ft./s
Atmospheric entry angle (with horizon):  6°
Reentry roll rate:  4.6 rps
Navigation Subsystem:  Mk 2 Mod 3 SINS binnacle and stable platform assembly, with Autonetics G7B Gyros.

[Platus (1986)]  [Francis, p. 161-162 (1995)]  [U.S. Patent #4,577,812 (1986), filed 1973]  [Buchonnet, p. 16, 23, 24 (1976)]

The W-68 yield of 40 kt, is the smallest strategic warhead ever fielded by the U.S. [NSC (1971)]  The Mk 3 RV AF&F (Arming, Fuzing, and Firing design "set new standards for miniaturization and provided improved protection against radiation".  [Loeber, p. 133 (2002)]

Might have had a beryllia (beryllium oxide, BeO) ceramic nose tip plasma arc torch sprayed ("plasma sprayed") with a beryllium surface coating.  [Be1018 (1969)]  [U.S.P.#3,791,851 (1974)]

Inaccurate, so generally thought as of a counter-value, rather than a counter-force weapon.  However in Congressional testimony in 1967, Paul Nitze stated that 10 x 50 kt [Poseidon] MIRVs would have the accuracy to destroy 1.2 to 1.7 hard target ICBM silos.  (a 84 - 99.5% chance of 10 MIRV destroying 1 to 2 hardened counter-force targets).  [Leitenberg, p. 4-5 (1984)]

Sprint ABM Missile (1974)

The high acceleration and velocity of this two-stage missile – a forty second time of launch-to-intercept -- required it have a warhead housing equivalent to an atmospheric reentry vehicle.  At its tremendous velocity, "the missile's skin became hotter than the interior of its rocket motor, and glowed incandescently." ["Seize the High Ground", Chapter 2]

The whole missile was of an aerodynamic conical shape, like a MIRV RV, with the sharp nose tip made of sintered tungsten metal infiltrated with silver or copper.  The silver, for instance, would boil off, cooling off the nose tip to the b.p. of silver. [U.S. Patent #3,883,096, Army (1975)]. The Sprint had nose tip had to resist a temperature of 6,200 °F, which is exactly the m.p. of tungsten. [Gibson, p. 210-213 (1996)] In fact, copper was used to infiltrate the tungsten metal.

The rest of the missile body used a silica fabric (Refrasil) -phenolic resin matrix ablative coating, a charring ablator.


Trident I (C4) SLBM (1979)

Mk 4 & Mk 4A RV's; 8 W-76 MIRV warheads; RV:  23.5" dia. x 7' high; 8° half-angle nose tip  [U.S.P. #4,185,558 (1980)]; higher ballistic coefficient than Poseidon RV.  Ballistic coefficient estimated at 1,800 lbs./sq.ft., compared to 2,000 for MM3 upgrade Mk 12A [Bunn (1984)]

42" wide woven graphite cloth impregnated with epoxy resin on a glass-reinforced phenolic honeycomb core, with integral graphite-epoxy (tape/cloth) end rings for the RV deck; solid propellant PBV; From rayon thread, woven into a cloth, and carbonized, the RV consisted of tape-wrapped carbon-phenolic (TWCP) bonded to a thin-wall aluminum substrate of the inner RV shell, with a fine-grained CMT graphite nose tip; the TWCP was similar to the Mk 12 RV material used in the AF MM3, but with the carbon particles eliminated.

With the graphite-epoxy material the fabric is woven from 3,000 filament T-300 tow, supplied by Union Carbide, which uses a polyacrylonitrile (PAN) precursor.

Used an AD3DX (commercial designation for a silicon body reinforced with a three-dimensional weave of fused silica fibers) antenna window.

New stellar-inertial guidance system.  [Spinardi, p. 127 (1994)]

Contractor:  Lockheed


Pershing II IRBM (December 1984)

RV:  welded aluminum monocoque (integrated vehicular) structure over-wrapped with a braided and pre-pregged tape consisting of rubber-modified, silica-phenolic (RMSP) [Refrasil, a charring ablator] and a glass-reinforced phenolic resin heat-shield.  The resin solids content was approximately 26%.

The low cost heat shield used a continuous/spliceless prepreg tape.  The simplest way to describe braiding is to illustrate a maypole dance with figures that represent braiding machine carriers.

A braided tape is defined by its shape (flat, in our case), the braid angle (the angle the yarns form with respect to the edge of the tape; 45°, 60°, 80°), yarn count (the number of yarn ends wound on each braiding machine carrier), and packing (which describes the fineness of the weave).

The total integrated heating was determined to be 4,375 BT/ft$^2$.  [NWDB-I, p. 297 (1984)] [AD-A113 928 (1982)]


MX/Peacekeeper ICBM (1986)

RV: Mk 21 ABRV (Advanced Ballistic Reentry Vehicle)

Dimensions: 0.55 m (21.8") dia. x 1.75 meters (68.9") high; Half-angle: 8.2°; Nose radius: 1.4".
Weight of graphite of RV: 125 kg.
The mid-bay section (which contains the warhead, but not the nose, or base) is 17.026" in diameter, and 33.10" high.
The structural shell consists of a basic T-300/5208 graphite/epoxy laminate, 0.055" thick, and with a 10 ply configuration. On the outside of the shell is a 0.005" thick layer of aluminum, bonded to the laminate beneath it. The mid-bay shell was built by Convair under contract with AVCO.

The ABRV consisted possibly as a one piece shell skin of laminated graphite composite (graphite/epoxy with judicious use of aluminum and Lexan plastic in suitable areas) for nuclear vulnerability hardening. [AD-A076 485 (1979)]

RV has a ballistic coefficient of 144,000 N/m2 (3,000 lb./ft2). Launched for reentry at 24° descent (angle with the horizontal), a minimum energy trajectory, it hits the ground with a speed of 3.4 km/s in a descent from 150,000 feet lasting 54 seconds, at a speed of 7 km/s. [Modern RV's travel at Mach 20 (7 km/s). [Aldridge, p. 92 (1983)] Its atmospheric friction heating rate maximizes at a 10,000 W/cm2 ["Countermeasures" (ca. 2000)] [AD-A231 552 (1990)].

Contractor: AVCO Corp. [Cochran, p. 126-127 (1984)]


Trident II SLBM (1989)

Mk 5 REB ABRV; RV: 22" base dia. x less than 6' high. [U.S.P. #4,623,106 (1986); Navy]

Carbon-carbon weave, supplemented by metal filaments along axis of symmetry, producing a more shape stable nose tip, and ablation more predictable, and subject to counteractions. [Garwin (1999)] [Spinardi, p. 153 (1994)] The Mk 500 REB had a carbon/carbon nose tip, and carbon/phenolic RV heatshield

Contractor: GE Co.; first SLBM RV not built by Lockheed.


## RV Hardening

Hardening are the measures taken to enable the RV and its warhead contents to withstand exposure to one or more effects of man-made hostile environments. "Hardness is the resistance or degree of resistance to nuclear blast countermeasures," the hostile environment, including neutron or x-ray exposure, or EMP microwave radiation, designed to disable, deflect, or destroy the incoming warhead.

"Brand X" is the insider's slang term used to describe hardening materials.

Neutron shielding is accomplished by a combination of three elements and was extremely difficult:

1) neutron moderation by light elements (Li6H, Be, BeO ceramic)
2) neutron reflection (BeO)
3) thermal or low energy neutron absorption (Cadmium, B10, Gadolinium)


Be and/or BeO are used as moderator or neutron-reflecting materials, and Li6H are used as shielding materials. [RDD-5, X.A.13 (1999)] Li7H would be useful for 14 MeV neutrons. Li6H and Boron-10 are used as a shielding material for neutron hardening of MIRVs [RDD-4, II-P-11 & V-A-2j (1998)]  [Wright (1997)].  They are both compressed to high density by HIP processing.

The Boron10 is used as a thin inner layer, surrounding the nuclear weapon casing.  Surrounding the B10 layer is the thicker Li6H layer which moderates the incoming high-energy neutrons, increasing the ability of the B10 layer to absorb and thus stop them.  The B10 is also mixed in as 10% of the LiH, being the most neutron absorbent material there is.  [AD-A206 093 (1989)]

Lithium-6 tetraborate was considered in 1960 for neutron hardening.  [Murray (1960)]  [DASA-1211 (1962)]

The DoE weapons labs are large consumers of BeO.  It is the best neutron reflector in existence, as well as a passable moderator.  BeO is also an excellent heat absorber, which deflects Spartan-type ABMs.  So BeO is used for neutron hardening of RV.  The BeO is prepared by sintering its powder in a high temperature isostatic pressing in a vacuum at 2,000 psi at 1,500 °C. by induction furnace.  [U.S.P. #3,279,917 (1966)]

Alternatively there is beryllium hydride, $BeH_2$ (density = 0.75 g/cm3), with 40% more hydrogen than LiH, used for moderating (hydrogen mostly) and reflecting neutrons.  [LA-1659 (1953)]

$10^{14}$ – $10^{15}$ neutrons/cm2 is a sure kill of an unhardened RV.  [AD-A115 691 (1982)]

Because of the great penetrative power of neutrons, x-ray shielding is easier than neutron shielding.  "The kill mechanism for x-rays is ablative shock."  [RDD-4, V-E-3g (1998)]  The kill radius for a 1 MT blast is 2 km for an unshielded RV.  [AD-A223 379 (1986)].  X-ray shielding (Be sheet covering a high-Z element, such as tungsten.  CCC (Carbon-Carbon Composite), LiH, and W were declassified as used as shielding material.  [RDD-5, X-A-13 (1999)]  CCC absorb or transmit high fluences of x-rays, depending on the radiation energy level.  The energy absorbed by the CCC is converted to heat, which is easily accomodated as a result of the composite's high specific heat and high sublimation temperature.  [AD-A325 314 (1996)]  High-Z materials are used in hardening MIRVs against high-energy x-rays.  [RDD-4, V-A-1i (1998)]

From further away, the x-rays would bathe the hollow plutonium core of an incoming MIRV's H-bomb primary, and melt or distort its shape by triggering at 115° C the alpha-phase plutonium to the beta-phase transition, rendering the Reentry Vehicle's bomb payload a dud.  [HW-79836, p. 3 (1963)]

Gamma ray (essentially much higher energy x-rays) hardening is impractical to shield against in an RV.  [CG-HR-1, Ch. 7 (1995)]

Lithium7 hydride (m.p. about 690 °C) has little competition with respect to heat absorption capacity.  Heating up to the dissociation point will absorb 143 kJ/mole, which equals 20.4 kJ/g (which equals 32.2 kcal/g), a tremendous amount.

As a complement to the hydride's ability to absorb heat, LiH is a very poor thermal conductor, which is an added bonus.  As far as x-ray opacity, it is here that LiH really shines as a useful material.  Additionally, Li7H has a very high effective neutron attenuation coefficient, and has an ability to capture high energy (i.e., H-bomb) neutrons without releasing gamma rays. [AD-A158 180 (1985)]

Since it reacts with water, low humidity and moist air, the required containers for LiH should be fabricated from 347 stainless steel.  Cold pressing of LiH can achieve theoretical density (0.775 g/cm3) at 15,000 to 30,000 psi for 30 minutes.  LiH needs an argon atmosphere during processing and use.  [AD-A314 441 (1996)]

BeO, the ceramic, is a better heat conductor than Be, lithium hydride, or pyrolytic carbon.  It also has a high melting point of 2,507 °C., and is useful as a heat sink, below the 0.5" thick pyrolytic carbon surface of the RV.  It is also used for RV x-ray hardening.

Tungsten (W) and uranium are the most efficient for gamma radiation attenuation, but the additional weight of these heavy metals militates against their use in RV's.

An x-ray hardened RV is sure-safe up to 70 cal/cm2, and a sure-kill at 100 cal/cm2. [AD-A115 691 (1982)] The 4 MT W-71 Spartan ABM warhead gives off 11 cal/cm2 at 50 km radius, enough to kill an unhardened RV. In comparison, a heat flux of 6 cal/cm2 is enough to be lethal to humans. [SAND91-0285 (1992)]

## Inertial Guidance Systems

> "One should not assess the power of missiles solely
> by...their throw weight. ...A more important criterion
> is the accuracy of the warhead. Doubling the accuracy
> is equivalent to increasing the warhead yield eight times."
>
> -- Soviet Marshal Sergei F. Akhromeyev (1985)
> quoted in "The Straw Giant" by Hadley

The 1960 Polaris A-1 was the first digital guidance system. A special purpose digital computer for guiding the flight of a ballistic missile, using the Q matrix guidance algorithm. It was the first digital computer INS ever used, replacing the previous analog systems with a great leap forward into the digital world. The hardware for the inertial system was 300 lbs. in 1960, but was reduced to 29 lbs. by 1970. [AD-A440 094 (1997)]

It is made up of a number of interconnected integrators that keep the position of the in flight missile known by determining velocity, and then position, by mathematically double-integrating the accelerometer inputs into the computer. Thus the computer can determine whether the missile needs course-correction, and how much.

The accelerometers are mounted on a three-dimension, gyroscopically-stabilized inertial platform. The accelerometers provided the inputs to the computer, whose output is a directional, corrective change to the engine motor's thrust direction.

The ideal gyroscope spins on a single vertical axis. What a gyroscope *actually* does is spin in two ways. Welcome to precession. The gyroscope body spins, but the gyroscope also spins in a spherical "wobble" (it actually moves in a perfect sphere) around the true vertical axis of the gyro body spin.

With [Draper's] incentive and drive, the Laboratory has over the years developed the necessary precision accelerometers and gyros and applied them to the inertial guidance of vehicles. In these systems, the gyros measure changes in vehicle direction or orientation; the accelerometers measure changes in vehicle velocity. The accelerometers sense these velocity changes in much the same way a blindfolded passenger senses the acceleration, braking, and turning of an automobile – but with much greater accuracy. However, a blindfolded passenger would be far less able to keep track of the changes in direction of motion than a gyroscope can in an inertial system. Since the accelerometers can measure only velocity changes and consequent position changes, the initial value of these parameters must be provided by some other source. Similarly, gyros can only sense changes in direction or orientation. Again, an initial value must be obtained by some alignment process. [Morgan (1998)]

The guidance system determines the missile accuracy. It determines by how far the warhead misses the target. The greater the accuracy, the smaller the yield, and thus the weight of the warhead. And with a lighter warhead, a smaller missile, or greater range with the same missile is achievable. Guidance systems are thus the lynch-pin of solid-fueled ballistic missiles.

The measurement of ballistic missile accuracy is known as CEP, "Circular Error Probable", which is the radius of the circle in which 50% of the warheads will hit inside. The CEP is particularly important for "hard target" kills: hardened ICBM silos, which require the greatest accuracy, or less preferably, yield.

The formula for calculation of kill probability is $(Yield)^{2/3}$ divided by $(CEP)^2$, with yield in kt, and CEP in meters.

Hard target kills were the essential goal of the fundamental "counter-force" nuclear strategy that was the goal of the U.S. since the early 1960s, and was finally achieved with the Minuteman 3 ICBM in 1980, and the Trident II SLBM in 1988. It replaced the counter-value strategy, where less accurate missiles could only target air bases, cities, and similar large, unhardened targets.

Counter-force requires an accuracy of 0.1 - 0.2 nm. [Lodal, p. 3 (1976)]

## The Evolution to the Advanced Inertial Reference Sphere – The AIRS FLIMBAL

The concept of inertial guidance made its debut with the 1940 MIT Ph.D. thesis of Walter Wrigley, one of Dr. Charles Draper's early students. Inertial guidance was demonstrated in February 1953, when an Air Force B-29 flew coast-to-coast, landing in Los Angeles, guided only by the INS, consisting of three gyroscopes and an analog computer. The total INS system weighed 2,700 lbs. The gyros measured changes in orientation and direction, and three accelerometers measured changes in velocity. An initial value oriented both devices to start the navigation process off. The first celestial, stellar INS system was demonstrated in 1949.

They were off and running.

In 1954, they began the system the evolved into future ballistic missile INS system, starting with the Thor, Atlas, Titan I and Titan II, and the Navy's Polaris onwards to Poseidon and Trident missiles.

> "In inertial-platform gimbal[led systems], the gyroscopes mounted on [the stable platform] measure angular rates, and gimbal-drive systems can use the angular-rate information to null the angular motion sensed by the gyroscopes. In this manner, the gyroscopes and accelerometers on the stable [platform] are inertially stabilized from the vehicle motion, and the stable platform physically represents an inertial reference frame. By double integration...of the indications from the accelerometers, with a correction for gravity [the missile position is determined.]" [AD-A055 778 (1978)]

Originated in 1957 by Philip Bowditch, who named it the FLIMBAL (Floating Liquid Gimbal), but was re-named the Advanced Inertial Reference Sphere (AIRS), is the third generation of INS, is the most accurate inertial navigation (INS) system ever developed, and marks the end of a lengthy, evolutionary process of continuous refinement of INS technology, begun by Dr. Charles Stark Draper of the MIT Instrumentation Laboratory, the leader in the development of hyper-accuracy gyroscopic INS.

The most novel aspect of the AIRS is that it has no gimbals. Gimbals are pivots that are provided for each of three spatial axes so that the guidance platform can move freely in all directions (and thus maintain its absolute alignment with the outside world). The AIRS consists of a beryllium sphere that floats in a fluorocarbon fluid within an outer shell and can thus rotate in any direction. The importance of this innovation is that it eliminates the possibility of gimbal lock (where the axes of two gimbals line up and destroy the three-dimensional freedom of motion), and is free from arbitrary limits to range of motion found in some gimbal designs.

The temperature of the fluid is controlled with extreme accuracy by transfer of heat from the fluid through "Power shells" to freon-cooled heat exchangers. The alignment of the sphere is controlled by three hydraulic thrust valves directed by the inertial sensors in the sphere. Like other INS systems, the sphere houses three accelerometers and three gyroscopes. The accelerometer design is called a SFIR (specific force integrating reciever). This is essentially the same approach as the Pendulous Integrating Gyroscopic Accelerometer (PIGA) used in the Minuteman II. The SFIR/PIGA works by measuring the rate of precession (and thus force applied) to a gyroscope at right angles to its axis of rotation. The gyroscope is a floated gas bearing gyroscope design. PIGA accelerometers use specialized technologies such as gas bearing wheels, ultra-stable ball bearings, precision electromagnetic components, and "designer chemical" flotation fluids. [AD-A386 542 (1998)]

The Advanced Inertial Reference System is used to measure the altitude of the vehicle with respect to an inertial reference. The AIRS platform consists of a hydraulically floated inner sphere inside an outer spherical case. The inner sphere contains the gyroscopes that are used to maintain the orientation of the inner ball relative to an inertial reference and the accelerometers which measure vehicle inertial accelerations and compute the velocity increments.

The inner sphere is maintained at an initial alignment relative to inertial space by a system of hydraulic jets that rotate the inner ball in response to signals from the gyros. The inertial ball has three printed circuit resolver "driver" bands which are mounted as three orthogonal great circles on the outer surface of the ball.

The platform attitude is measured in terms of the intersection points of the driver bands with a "receiver" band, also mounted as a great circle on the inner surface of the case. The positions of these intersections are defined in terms of the angles X1, X2, X3 measured along the receiver band and in terms of the angles $\varphi1$, $\varphi2$, $\varphi3$ measured along the driver bands.

The AIRS programs are usedf to extrapolate and interpolate the AIRS band angles and compute the transformation matrices. The AIRS programs receive the platform-to-body transformation matrix and the values of receiver and driver band noise from the main program and compute quantization and deterministic (table look-up) errors to determine the measured incremental altitude vector and the measured body-to-platform transformation matrix. [AD-A058 574 (1979)]

The AIRS is the "third generation" of INS, achieving drift rates of less than $1.5 \times 10^{-5}$ degrees per operational hour. This drift rate is so low that the AIRS contributes on the order of only 1% of the Peacekeeper missile's inaccuracy, and is thus effectively a perfect guidance system (i.e. a zero drift rate would not measurably improve the Peacekeeper's performance.

Very little of the precision of this guidance system is even exploited during a ballistic missile flight, it mostly being used simply to maintain guidance system alignment on the ground during missile alert, without needing an external reference through precision gyro-compassing. Most ICBMs require an external alignment system to keep the INS in synch with the outside world prior to launch. The AIRS is probably as good as an INS for ICBM guidance needs to get.

The cost for this extreme level of accuracy is, not surprisingly, of tremendous complexity and expense. For example, the AIRS has 19,000 separate parts. In 1989, just one of the three accelerometers used in the AIRS, cost $300,000 and took six months to manufacture.

There are very few applications beyond ICBM guidance, requiring both such precise guidance and independence from external references. If the requirement for complete autonomy is eliminated, extreme guidance accuracy is available at a small fraction of its cost and weith, but at the cost of indepence from external interference and jammability.

For example, satellite positioning systems like GPS (Global Positioning System) and GLONASS, permit centimeter level accuracy over unlimited periods of operation with only a light inexpensive receiver. NASA spacecraft require extreme guidance precision, but uses external navigation cues to obtain it.

The MX (later renamed the Peacekeeper) missile began development in February 1972. The military requirements for this missile called for greatly enhanced accuracy that the AIRS was suitable to deliver. In May 1975 the AIRS was transferred from the Draper Laboratory to Northrop for advanced development. It proved extremely difficult to transfer the hand-crafted laboratory built design to a production environment. Despite years of work, by July 1987 Northrop Electronics Division had succeeded in delivering only a small number of INS units. MX missiles were beginning to pile up in silos with no guidance system to fly them. By December 1988 AIRS units had been supplied to all 50 MX missiles. Since then, the contract by producing AIRS unit had been transferred to Autonetics Division, Rockwell International.

With the planned retirement of the Peacekeeper missile under the terms of the Start II treaty (signed January 3, 1993) by 2004, the Minuteman III was left as the sole U.S. land-based ICBM through the year 2020. Because of this shift in importance, the Minuteman III force is to be upgraded to the same standards of accuracy of the Peacekeeper. Accordingly between 1998 and 2002, 652 new AIRS guidance units were purchased and retrofitted to the existing Minuteman III's.

The most novel aspect of the AIRS is that it has no gimbals. Gimbals are pivots that are provided for each of three spatial axes so that the guidance platform can move freely in all directions (and thus maintain its independent alignment from the outside world). AIRS consists of a beryllium sphere that floats in a fluorocarbon fluid within an outer shell and can thus rotate in any direction. The importance of this innovation is that it eliminates the possibility of "gimbal lock" (where the axes of two gimbals line up and destroy the three-dimensional freedom of motion), and is free from arbitrary limits to the range of motion found in some gimbal designs.

The temperature of the fluid is controlled with extreme accuracy by transfer of heat from the fluid through "Power Shells" to Freon-cooled heat exchangers. The alignment of the sphere is controlled by three hydraulic thrust valves directed by the inertial sensors in the sphere.

Like other INS systems, the sphere houses three accelerometers and three gyroscopes. The accelerometer design is called a SFIR (specific force integrating receiver). This is essentially the same approach as the pendulous integrating gyro accelerometer (PIGA) used in the Minuteman II. The SFIR/PIGA works by measuring the rate of precession (and thus force applied) to a gyroscope at right angles to its axis of rotation. The gyroscope is a floated gas bearing gyroscope design.

The accelerometers and gyroscopes are descendants of technologies used in earlier ICBM INS systems like the Minuteman II. These technologies were developed over a period of 30 years by the Charles Stark Draper Laboratory (formerly called the Instrumentation Laboratory of MIT).

The gimballess floated sphere was conceived at the Instrumentation Laboratory in 1957 by Philip Bowditch, who dubbed the concept the "FLIMBAL" (Floating Inertial Measurement Ball). It was developed into a deployable system by Kenneth Fertig, under an Air Force program known as SABRE. In 1969 the highly accurate ICBM guidance program was eliminated, but resurrected as the MPMS (Missile Position Measurement System). It was test flown riding "piggy back" on a Minuteman III in 1976 (i.e., in addition to the actual Minuteman III NS20 guidance system.

We live in 3-dimensional space. It can be described mathematically by three flat planes, what mathematicians characterize as the X-Y-Z planes, with X being, say your North-South location, Y being East-West, and Z being your distance above the earth's surface (or more precisely your height above sea level. Ships use just the X-Y plane to navigate, which are known as latitude and longitude.

For airplanes and rockets, there is the equivalent 3 axes for the vehicle's movements in one place on its trajectory, called pitch, yaw, and roll. Pitch (nose up or down), yaw (nose left or right), and roll (rotation in a circular fashion, either to the left or right – a spinning ICBM is in an uncontrollable roll; though before inertial guidance systems were developed, rapid spinning was used for gyroscopic stabilized flight to keep the rocket on a straight line flight path).

The missile guidance set (MGS) is the system which continually determines the position in space, including altitude and position, velocity, and acceleration (all directly mathematically related to one another) of the vehicle, and adjusts the velocity and direction of the vehicle if it is in an incorrect position, according to a stored value of the pre-computed trajectory to the target. The trajectory is the sequence of positions in space at the various times after launch, until the target is reached. And the MGS is completely "self-contained". In other word, it works independently of any outside influence, not like radio command guidance.

The central part of Inertial Guidance System is the inertial reference and velocity measurement unit (IMU) consisting of a platform that is stable in three dimensions, using 3 gimbals or "frames" or "frame gimbals", each inside of one another. This platform isolates it from the missile by using three high accuracy gyroscopes, and carrying three PIGA accelerometers to take the measurements of acceleration, which is then mathematically integrated to give the new velocity.

Thus the MGS determines the missile's positional change from its launch point (input by the missile crew) by determining any changes in the missile's direction, and any changes in velocity.

There is in more modern IMU's a fourth gimbal, added to bring back into stability the platform after a sudden violent change in position of the missile, and to eliminate gimbal lock.

The IMU's output outputs its data to a "torque motor", which in turn helps correct the missile thrust and direction to correct for any indicated deviance ("off course" position) from the intended trajectory.

A complete literature and patent survey of the missile gyroscope has been completed [AD-A197 162, Part 1] [AD-A221 593, Part 3], including modern gyroscope construction: floated gyros, gas bearing gyros, electrostatic gyros, magnetically supported gyros, et al.

Cobalt-samarium ($SmCo_6$) was developed in the late 1960's, and is the best magnetic material of any developed, with a coercive force of 21,700 Oersted, ten times greater than any other (such as Alnico – aluminum/nickel/cobalt), and energy products two or three times greater. In 1968, they found that coating the SmCo6 particles with zinc or tin improved performance. [AD-A528 970 (2005)]

The Missile Guidance Computer (MGC) feeds the accelerometer reading into integrators to mathematically integrate the acceleration to give the vehicle's velocity (Calculus! Remember?), and then integrates the velocity measurement to give the vehicle's position relative to its original launch point.

The Bryant Grinder Corporation, Vermont, is a high-precision ball bearing grinder, using their "Centalign-B" machine, to grind to 25 µinch accuracy micro-ball bearings for three gimbal, high-speed gyroscopes.

The INS does for geometry – angles, speed and distance – what a watch does for time, Charles Stark Draper once said. (And a sextant did for latitude measurement before LORAN radio navigation, and then GPS satellite navigation.) Draper was a professor at MIT, and founder of the Draper Instrumentation Lab at MIT. He took Walter Wrigley's idea of inertial guidance and ran with it, becoming the highly productive and respected Dean of INS for 25 years.

The first INS used three single-degree-of-freedom gyroscopes to establish an inertial reference coordinate system. An on-board computer was then used to transform the navigation state from inertial coordinates to an earth-centered, geodetic coordinate frame for navigation purposes. [AD-A108 511 (1980)] The gyros measure changes in missile direction and orientation in three-dimensional space. The accelerometers measure changes in vehicle velocity. The initial position must be manually punched in.

The INS accelerometer has always been succeeding generations of PIGA, whose origins date back to the WW2 V-2 SRBM. PIGA sub-components include gas bearing wheels, ultra-stable ball bearings, and special flotation fluids.

In a missile the INS manages the direction and duration of rocket burn, in the initial boost phase of powered flight which lasts only 5 - 10 minutes, so that the final 20 - 25 minute unpropelled coasting phase of the payload is left on the desired trajectory required. Unlike radio command, which was susceptible to interference or jamming, INS is independent of any external influences.

To eliminate the three-axis mechanical supports, called gimbals, and are a source of inaccuracy, the Draper Lab developed the "beryllium baby", a floating precision-machined beryllium metal floating, centered in a close-fitting outer shell, which provides a "womb-like" environment for the inertial sensors. The new design was called FLIMBAL, a sub-component of the AIRS, and was the Lab's third generation INS. [U.S.P. #4,035,762 (1977)

The beryllium sphere is manufactured by sintering finely powdered beryllium metal, which is compacted under vacuum with an isostatic press, while being heated to below its melting point. [U.S.P. #3,880,606 (1975)] The beryllium sphere is approximately 12" in diameter, and floats in a fluorocarbon (Freon-like) liquid of high density such that the sphere is neutrally buoyant (about 1.84 g/cm3, the density of beryllium metal). [Drake (1967)] The flotation fluid must also have a high viscosity for good motion damping properties. Examples of such fluids are chlorotrifluoroethylene (CTFE), or bromotrifluoroethylene (BTFE). [U.S.P. #4,835,304 (1989)]


## Q-Guidance

Finally, there was the software implementation of guidance running on the on-board computer.

In the summer of 1952, Dr. Richard Battin and Dr. J. Halcombe ("Hal") Laning Jr., researched computational based solutions to guidance, as computing began to step out of the analog world and into the much superior digital age.

Laning, with the help of Phil Hankins and Charlie Werner, initiated work on MAC, an algebraic programming language for the IBM 650, which was completed by early spring of 1958. MAC became the work-house of the MIT Lab. MAC is an extremely readable language having a three- line format, vector-matrix notations and mnemonic and indexed subscripts.

Laning and Batin did the initial analytical work on the Atlas inertial guidance in 1954. Other key figures at Convair were Charlie Bossart, the Chief Engineer, and Walter Schweidetzky, head of the guidance group. Schweidetzky had worked with Wernher von Braun at Peenemuende during World War II.

The Q-system was presented at the first Technical Symposium on Ballistic Missiles held at the Ramo-Woodridge Corporation in Los Angeles on June 21 and 22, 1956. Derivations of Q-guidance are used for today's military missiles.

Q-guidance is used for missiles whose trajectory consists of a relatively short boost phase (or powered phase) during which the missile's propulsion system operates, followed by a ballistic phase during which the missile coasts to its target only under the influence of gravity. The objective of Q-guidance is to hit a specified target at a specified time (the target moves with the earth's rotation.

The initial "Delta"-guidance system (that was the predecessor to Q-guidance) assessed the difference in position from a pre-planned, pre-calculated reference trajectory, done by large computers on the ground, ahead of time. Using these ground-based computers, a reference trajectory is plotted well before the flight and stored in the MGS computer.

In flight, the actual trajectory is modeled mathematically as a Taylor series expansion around the reference trajectory. The guidance system attempts to zero the linear terms of this expansion in order to bring the missile back to its reference trajectory.

In contrast, Q-guidance is a dynamic method, reminiscent of the theories dynamic programming or state-based feedback. A velocity to be gained ($V_{GO}$) calculation is made to correct the current trajectory with the objective of driving $V_{GO}$ to zero, when the pre-planned trajectory is regained.

The mathematics of this approach were fundamentally valid, but dropped because of the challenges in accurate inertial navigation (eg. IMU accuracy) and analog computing power.

The challenges faced by the Delta-guidance efforts were overcome by the "Q system" of guidance, introduced in June 1956 at a classified technical symposium held at the Ramo-Woodridge Corporation in Los Angeles.

The Q system is a dynamic system. "Dynamic" means that it is not linear, but time-varying or position-varying. Dynamics is the mathematics of varying change, such as can be found in velocity or acceleration.

The Q system involved not maintaining a pre-defined trajectory, but calculating the velocity to be gained to set a new course to hit the target at the required time. This was more efficient than correcting to a pre-set course.

In the Q system, the matrix Q represents the partial derivatives of the velocity with respect to the position vector. The components of the vector cross-product (v, xdv/dt) are used as the basic autopilot rate signal, a technique which became known as "cross-product steering".

But now a quick initial divergence to mathematical definitions.

In math, there are scalar quantities and vectors. Scalars are thing like mass, volume and density and are associated with just one number (thankfully!). A vector, on the other hand, has both magnitude and direction. Examples are velocity, acceleration and force. (Though they are commonly used interchangeably, speed is scalar and velocity is the vector.

A vector is represented mathematically by three numbers (x,y,z) associated with it, corresponding to the three dimensions of space, and this complicates the math of "vector mechanics" for the uninitiated. A vector is denoted in print by a letter in **bold**. There are some operations connected with vectors. One of them is (**R X S**), which is known as the cross-product of vectors **R** and **S**. The result of this is a vector **T**, of magnitude (RS sin α), and whose direction is orthogonal (90°) from the plane formed by **R** and **S**. α is the angle between **R** and **S**. R and S (**not** in bold) are how the magnitude of these vectors is shown.

The Q system's revolution was to bind the challenges of missile guidance (and associated equations of motion) into a matrix Q. The Q matrix represents the partial derivatives of the velocity, which gives the missile position. A key feature of this approach allowed for the components of the vector cross product (v, xdv/dt) to be used as the basic autopilot rate signals. This technique became know as "cross-product steering".

In essence Q-guidance says "never mind where we were supposed to be, given where we are what should we do to make grogress towards the goal of reaching the desired target at the required time." To do this relies on the concept of "velocity to be gained." At a given time, t, and for a given vehicle position r, the correlated velocity vector $V_c$ is defined as follows: if the vehicle had the velocity $V_c$ and the propulsion system was turned off, then the missile would reach the desired target at the desired time under the influence of gravity. In some sense $V_c$ is the desired velocity. The actual velocity of the missile is denoted by $V_m$, with the missile subject to only the acceleration due to gravity, **g**, and that due to the thrust of the engines, $a_T$. $V_{TBG}$, the velocity to be gained, is defined as the difference between $V_c$ and $V_m$ :

$$V_{TBG} = V_c - V_m$$

A simple guidance strategy is to apply acceleration (i.e., engine thrust) in the direction of $V_{TBG}$. This will have the effect of making the actual velocity come closer to $V_c$. When they become equal (i.e., when $V_{TBG}$ becomes zero) it is time to shut off the engines, since the missile is by definition able to reach the desired target at the desired time coasting on its own.

The only remaining issue is how to compute $\mathbf{V_{TBG}}$ easily from information available on board the vehicle. A remarkably simple differential equation (starting with the derivative with respect to time of $\mathbf{V_{TBG}}$) can be used to compute the velocity to be gained:

$$dV_{TBG} / dt = -\alpha T - Q\, V_{TBG}$$

where Q is a symmetric 3 by 3 time-varying matrix defined by:

$$Q = \delta V_c / \delta r \,|\, r_T, t_f$$

(The vertical bar refers to the fact that the derivative must be evaluated for a given target position $r_T$ and time of free flight $t_f$ ). The calculation of this matrix is not trivial, but can be performed offline before the flight, experience shows that the matrix is only slowly time-varying, so only a few values of Q corresponding to different times during the flight need to be stored on board the missile. In early applications the integration of the differential equation was performed using analog hardware, rather than a digital computer. Information about vehicle acceleration, velocity and position is supplied by the on-board IMU.

## Derivation of the Equation

Notation:  t the current time, r the current vehicle position vector, $\mathbf{V_m}$ the current vehicle velocity, vector $\mathbf{T}$ the time the vehicle will reach the target, $t_f$ the time of free flight for the correlated vehicle, i.e., T - T. A reasonable strategy to gradually align the thrust vector to the $\mathbf{V_{TBG}}$ vector is to steer at a rate proportional to the cross product between them. A simple control strategy that does this is to steer at the rate:

$$\omega = \kappa\, (\alpha T \ \times \mathbf{V_{TBG}})$$

where κ is a constant. This implicitly assumes that $\mathbf{V_{TBG}}$ remains roughly constant during the maneuver. A somewhat more clever strategy can be designed that takes into account the rate of time change of $\mathbf{V_{TBG}}$ as well, since this is available from the differential equation above. This second control strategy is based on Battin's insight that, "If you want to drive a vector to zero, it is [expedient] to align the time rate of change of the vector with the vector itself." This suggests setting the auto-pilot steering rate to:

$$\omega = \kappa\, ( \mathbf{V_{TBG}} \ \times \ d\mathbf{V_{TBG}} /dt )$$

Either of these methods are referred to as cross-product steering, and they are easy to implement in analog hardware. Finally, when all the components of $\mathbf{V_{TBG}}$ are small, the order to cut-off engine power can be given. [wikipedia.org (2012)]

## SLBM Launching

In the case of submarine INS, the system provides continuous accurate measures of vehicle position, orientation and velocity under the effect of manual propulsion and steering, and random water and wind motion.

The first submarine navigation system was SINS, the Ship INS, developed in 1954

The first stellar INS, that used stars as reference points to improve navigational accuracy, was demonstrated in 1949. It was finally implemented and used in the Trident I to supplement its regular INS. [Morgan (1998)]

The Stellar Inertial Guidance System (SIGS) utilizes a stellar sighting to correct for errors in the INS from inertial component or geodetic and geophysical error sources. The INS digital computer needs an addition RAM memory size of 1,500 words to contain the SIGS program requirements. [Drake (1967)]

Stellar guidance has a limited catalog of usable stars and "orbital elements" (the planets Venus, Mars, Jupiter) and associated position data that the stellar sensor can reference itself too. The position data constants are time-dependent. Position constants vary with proper motion, precession, nutation, aberration, and heliocentric parallax.

A description of an older (1977) stellar system is as follows. The key element of the Northrop NAS-26 system, the astro-inertial instrument, is a three gimbal stable reference platform with an integral two degree of freedom star tracker. The reference platform contains two two-degree of freedom gyros and three accelerometers. The system weighed only 184 lbs.

The star tracker, which is highly accurate, provides correction of inertial platform errors by precisely tracking an average of three stars per minute, day or night, at any altitude. The system tracks down +3.5 magnitude stars and bright stars in backgrounds up to 8,000 foot-Lamberts. The stellar ephemeris provides a minimum of two star availability, world wide, with six to ten stars typically available.

The TTL MSI (transistor-to-transistor logic, medium scale integration) computer has a clock speed of 4.5 MHz, with a 65k 16-bit word magnetic core memory [AD-A090 649 (1980)]

The stellar sighting is used to calculate the INS error. These errors are calculated using a pre-computed 9 x 2 gain (or weighting) matrix to the selected star sighting. The computation of the gain required to match the supposed position of the star, leading to an estimate of the guidance platform misorientation. Thus can be calculated the position, velocity, and orientation errors.
[AD-A090 649 (1980)]

The ideal rotor for an electrostatically suspended gyroscope is a very nearly perfect sphere, relatively light in weight and stiff as possible. Beryllium is thus ideal, being the stiffest metal of all.

| ICBM: no uncertain points | Vs. | SLBM: one uncertain point (the launch point); more |
|---|---|---|
| accurate | | or less accurate |
| ICBM: further range | | shorter range |
| vulnerable; non-survivable | | survivable |

Accelerometer (1947, declassified in 1952) (U.S.P. #2,613,071)
Accelerometer (1957, declassified in 1973; Sperry Rand) (U.S.P. #3,713,343) is the basic patent for
        accelerometers, with a spherical weight suspended by a spring, with takeoff springs in
        the three dimensions measuring acceleration.
Accelerometer (1959, declassified in 1975; Rockwell International) (U.S.P. #3,863,508)
Accelerometer (1964, declassified in 1971; North American Rockwell) (U.S.P. #3,597,598)

Radiation-hardened amplifier for thin film memory [U.S.P. #3,969,737(1975) (filed 1969; classified 6 years)

** Polaris Guidance System, U.S.P. #4,405,985 (filed 1965, declassified 1983). Eldon C. Hall et al.
** Polaris Guidance System, U.S.P. #4,470,562 (filed 1965, declassified 1983). Eldon C. Hall et al. See
        also U.S.P. #2,932,467 (filed 1955, declassified 1960), and B.P. #1,605,273
        (filed 1965, declassified 1987).

U.S.P. #3,990,657, and U.S.P. #5,442,560 (1995).

*   *   *   *

Thor IRBM (1958)

CEP:  1 km  [Sheehan (2009)]
      > 2 nm


Jupiter IRBM (1958)

Inertial-guidance:   gyro-stabilized platform (ST-90) mounted with three air-bearing supported accelerometers connected to a separate computer which received its input and calculated distance and speed; heavily based on Redstone system.

CEP:  1.5 km  [Grimwood (1962)]


Atlas ICBM (1959)

CEP:  3.8 km (2 nm)  [Goldberg, part 2, p. 872 (1981)]


Polaris A-1 SLBM (1960)

Initial ball bearings of early designs replaced by gas spin bearings with a substantial improvement in performance of the Mk1 guidance system for the Polaris A-1.   Even a 4 mile CEP would have been considered acceptable, and was met "as far as we could tell" [Spinardi, p. 59 (1994)]  PIGAs – pendulous integrating gyroscopic accelerometers – with gyro wheel and internal gimbal.

For the Polaris A-3 MRV, the Mk 1 was upgraded to the Mk 2 with smaller spherical aluminum gimbals replacing the large bar beryllium gimbals of the Mk 1, reducing the weight one-third and the size to one half.  PIGAs replaced by smaller (1.5" diameter) PIPAs – pulsed integrating pendulous accelerometers – on all but the most sensitive thrust axis.  Consists of a pendulum held in null position by pulses sent from a signal generator to an electromagnetic torquer.  Simpler, and cheaper than PIGA, but does not match the sensitivity of the best PIGAs.  [Spinardi, p.70 (1994)]

To save weight, magnesium was chosen for the stable member on which the inertial components were mounted, and "cordwood" welding used instead of soldering for the guidance electronics packaging.

Also, for the Polaris A3, the SINS was upgraded to the Mk 2Mod4.

By 1964, the APL (Applied Physics Lab) of Johns Hopkins University had developed a sophisticated model of the gravitational field of the earth, based on a variety of APL satellite measurements.


Titan I ICBM (1960)

CEP:  1.6 km
CEP spec of 1.2 km or less.  Actual CEP during 1965 testing was 1.5 – 1.7 km (0.7 nm – 0.8 nm)

The guidance system continuously determines the precise missile position and from this data determines missile velocities in three coordinates. A stable reference along the pitch and yaw axis is provided by reference gyros in the flight control system. Deviations from the established reference cause the thrust chambers of the missile rocket engines to gimbal to correct the missile altitude. [T.O. 21-SM68-1 (1964)]

Minuteman I (1962)

The MM1 had a NS10 Missile Guidance Set (MGS) [Reed (1990)] [Adkins], consisting of a gyro-stabilized platform, a guidance computer, and memory unit.

Guidance System NS10Q: used a D-17B general purpose digital mini-computer, shaped like a right cylinder maximum diameter of 29", 20" high, 5" in depth, and weighing 62 lbs. It had 76 printed circuit boards and a hard disk memory unit. Its power supply consumption is around 700 w. It also included a gyro-stabilized platform, and a forced air cooling system to dissipate the computer's 350 watt power consumption. The power supply needs an input of 28 VDC and 19 - 25 amps. Its high reliability was achieved by using DRL (Diode-Resistor Logic) extensively, with some DTL (Diode-Transistor Logic) where necessary.

The D-17B was made by Autonetics in 1962, and was a 24 bit word computer (with an additional 3 timing bits added) with a clock frequency of just 350 kHz, 39 instructions in the instruction set, and 2,800 word (24 bit word) hard disk. There were 48 digital output lines and 28 digital input lines, 12 analog lines, and 3 pulse lines. The memory addressing was direct. The number system was binary, fixed point, sign plus 2's complement.

The electronics had a liquid cooling system. [AD-746 008 (1972)] There were 1,521 transistors, 6,282 diodes, 1,116 capacitors, and 5,024 resistors, mounted on double-sided, gold plated, glass fiber laminate printed circuit boards. [AD-760 757 (1973)] The cost was $250,000 in 1962 dollars.

Punched tape was used to load program and input data.

Sergeant SRBM (1962)

"The range, or distance, of the ballistic trajectory of the *Sergeant* solid-propellant surface-to-surface guided missile is controlled primarily by an aerodynamic dragbrake which operates three times during the course of flight. The first braking period provides gross adjustment of range; the two remaining braking periods provide correction for nonstandard performance and atmospheric conditions. The dragbrake consists of four flat-faced blades and an actuating mechanism to extend the blades into the airstream. The missile drag with blades extended is approximately five times as great as the drag with blades retracted. The brake is actuated in a plane normal to the missile axis. The actuation time for extending or retracting the blades is 1.25 sec." [Thorman (1959)]

Pershing I/IA IRBM (1964)

The Pershing I used a similar system as the The Pershing I RV used a miniature (about 10 lbs. in weight) radar altimeter, to determine height, and incorporating a transistorized crystal-controlled time base, a pulse modulator, a time interval computer, a high power modulator, and a miniaturized oscillator cavity. The magnetron produced a 4 mm wavelength (70 GHz) signal of power 0.1 kW, and was fed its signal via a microwave stripline (a photo-etched flat copper, coaxial, possibly Teflon, printed circuit board, first used in the Little John SRBM). [AD315 467 (1959)]

Minuteman II ICBM (1966)

Used a NS17 MGS. It used beryllium gimbal rings for the gyroscope, and contained PIGA accelerometers. [AD0477 517(1965)]

Guidance System NS17: Contains a gyro-stabilized platform with sub-components of the G6B4 and GI-TI-B gyroscopes, and Gyro-compass sub assembly. On it is the PIGA units. Data is processed by the D37C MGC (Missile Guidance Computer) computer, using nuclear-hardened thin film memory [U.S.P. #3,909,739 (1975)] with hard disk, which contains the MM3 codes and SIOP data.

The D-17B computer had its limitations. Its main memory size was relatively a very small memory size. In addition, the memory access time was comparatively long, since it had a serial access disk memory, rather than the much faster random access memory (RAM).

The D-37C MGC which was an upgrade in 1964 over the D-17B, was essentially a micro-miniaturized, upgraded version of the D-17B. The memory capacity was increased about 3.6 times. A more varied instruction set of 58 (versus 39 for the D-17B) was created for more versatility. And ICs (integrated circuits) were used for the first time, replacing the much larger transistors, and other discrete components used in the now obsolete D-17B.

D-37C was a 24 bit word computer with a 7,200 word hard disk. The D-37C was used in the Minuteman II and III, initially. Input was by running paper or Mylar tape through a reader, and consisted of a 57 word instruction set. Like the D-17B was built by Autonetics of North American Rockwell and had the same clock speed. It was built with small scale ICs, using diode-transistor logic, which resulted in a size reduction over the D-17B's individual transistors and other components of 21" x 7" x 9.5", and a price of $139,000 in 1964 dollars.

Examples of the basic instruction set include:

| Mnemonic Code | Description | Numeric Code |
| --- | --- | --- |
| ADD | Add | 64 |
| ALC | Accumulator Left Cycle | 00 |
| CLA | Clear And Add | 44 |
| COA | Character Output A | 00 |
| COM | Complement | 40 46 |
| DIA | Discrete Input A | 40 02 |
| EFC | Enable Fine Countdown | 40 26 |

[AD777 244 (1974)]

Minuteman III ICBM (1970)

The basis of the MM3 INS is the gyro-stabilized platform, which besides providing acceleration (and therefore velocity) and attitude information to the MGC, also provides level detector and gyro-compass information, and accepts control signals from the MGC so that platform attitude constants are obtained and then the platform is properly aligned prior to missile launch.

The GPS carries three PIGAs to measure missile acceleration along all three axes in 3-D space, and uses an external gimbal configuration. The platform is stabilized by two dual-axis, free-rotor gyros whose rotors are supported on self-generated gas bearings. One gyro serves as the pitch and roll axis stabilization reference, while the other provides an azimuth stabilization reference.

Each PIGA accelerometer contains a gyroscopic pendulous mass which is floated in a liquid to minimize bearing load and friction. ["Minuteman Weapon System" (2001)]

Like the MM II, the MM III originally used a NS17 MGS.

The MM3 was upgraded in 1976-1979 from the NS17 to the NS20 MGS, along with its emplacement in the new MX/Peacekeeper ICBM. In the MM3, a silver-zinc battery provided 28 volts dc for in flight operation of the Flight Control Systems, the NS20MGS, and the Reentry System.

NS20: Contains a gyro-stabilized platform, with sub-components G6B4 and GI-TI-B gyroscopes, Gyro-compass sub assembly, and coolant lines. On it is 16 PIGA Mod G accelerometer units. Data is processed by first the D37C computer, then upgraded to the D37D computer with hard disk. There is a P92 Amplifier Assembly, and an SE437A battery.

NS50: Scheduled for installation in the mid to late 2000's. It has the same components: G6B4, G1-TI-B gyros. The MGC uses battery-backed up SRAM memory, and is equipped with a high-speed 16-bit microprocessor.

Geodetic satellite correction for gravity anomaly course overflight correction, and addition of stellar-tracking system.

Poseidon SLBM (1971)

This next iteration of the FBM used the Mk 3 guidance system upgrade using three 1.6" diameter PIPAs, using permanent magnet torquers. It was controlled by a general purpose computer with 100k of ROM and 12k of a type of RAM called plated wire memory. The plated wire memory was chosen because it was radiation hardened – the only one available.

Lance SRBM (1972)

CEP: 375 m at 125 km range

Guidance: The simplified, strap-on ("body-fixed") gyroscopic inertial guidance system is actually more accurate than field artillery; the Lance is a close support weapon for troops. The guidance set weighs less than 40 lbs. and being a strap-on gyroscopic system is less accurate, but also more versatile, less complex, and cheaper than regular, more accurate ICBM gyroscopic inertial guidance systems. Since the Lance is a SRBM, the loss of accuracy is small enough that it is not important.

The guidance set is made up of three subsystems: the directional control (DC) subsystem, the velocity control (VC) subsystem, and the power supply subsystem.

By regulating the four TVC valves, the DC subsystem maintains correct missile attitude about the pitch and yaw axis during the boost portion of flight. The major components of the DC subsystem are: a two-degree-of-freedom RSG-15 gyroscope and a directional control electronics assembly. The precisely machined inside of the gyroscope rotor housing supports the rotor through a spherical bearing. The rotor housing has a stiffening end plate, which mounts to the housing with 12 screws. It was added to provide axial stiffness to the rotor support plane. There is also mounted internally an electrical pickoff system.

The VC subsystem which contains a 4841 Accelerometer and a velocity control electronics assembly, measures missile velocity and shuts off the booster when engine missile velocity reaches a preset value. After boost termination, the VC subsystem maintains missile thrust equal to drag by controlling sustainer engine operation.

The Lance uses 3 Q-Flex (Quartz Flexure) accelerometers. Unlike the gyroscope, the accelerometer does have significant internal electronics. The Q-Flex accelerometer is a pendulous accelerometer of the torque-rebalance type. It consists of a pendulous/proof mass, pendulously supported, and constrained so as to allow only one degree-of-freedom about a well-defined axis (one of the three axes, x,y, and z) fixed within the sensor.

The basic Q-Flex accelerometer is made from two major sub-assemblies, the Q-Flex sensor, and the hybrid servo electronics. The Q-Flex sensor contains the acceleration sensitive element. It is a linear, single axis, electro-mechanical device for measuring acceleration.

The torque required to maintain the pendulous mass in a position that is fixed relative to the accelerating missile in which it is mounted is a direct measure of the acceleration. The torque is converted to a voltage or current measurement for use by the rest of the guidance system.

The Q-Flex accelerometer uses a quartz flexure for suspension of its proof mass to avoid the inelastic hysteresis of metal flexures (i.e. the quartz flexes a constant amount reliably and lastingly, unlike metal). The proof mass and flexures in a Q-Flex accelerometer are formed from a single blank of specially processed fused silica. [AD-A121 264 (1982)]

There are 2 torque coil aluminum (then later replaced by quartz) bobbins in the Q-flex sensor which are attached to the quartz reed with an adhesive. Any stress on the reed will cause it to warp and produce a bias error through the action of the servo electronics. [AD-A063 901 (1978)]

The guidance set also includes a timer to fire the squibs that close the torque-termination valve when the desired missile spin rate is achieved.

The Lance missile has three phases of flight: boost phase, sustain phase, and coast phase. In boost phase, the missile is accelerated to the required velocity. The correct heading/direction was set at launch, and is maintained by proper directional control. Next, in sustain phase, the missile reduces all external forces except gravity to longitudinal drag and provides the proper amount of thrust to cancel it. Sustainer engine cutoff begins the coast phase.

By launching at a specific launch angle (54° for long ranges – greater than 90% of the maximum range – and 48°, for less than 90% of the maximum range), and by specifying the velocity reached at boost phase termination, the range can be figured out in advance of launch.

The performance requirements for the Lance accelerometer are more stringent than for the gyro since it must function accurately at accelerations up to 30 g's during boost phase. [AD-A017 242 (1975)]

Inertial-guidance by Kearfott Division of General Precision Inc.; Guidance info is fed to the SRAM by the B-52's inertial-guidance system, designed by Litton Industries.

Spartan ABM (1975)

Was equipped with the FLIMBAL gyro for improved accuracy, along with a missile guidance set designed to deal with "severe shock, vibration, and noise".

Trident I SLBM (1979)

The next FBM, Trident I had a new Mk 5 guidance system, which included a new star sensor, the Kearfott Unistar stellar-inertial system, mounted on the gyro-stabilized platform, along with the gyros and accelerometers. The GSP was held in a four gimbal system, rather than the three used in Poseidon and Polaris. The extra gimbal was used for optical alignment with the SINS, and other gimbal devoted to elevation of the star sensor through the vertical plane of the predicted star's location.

Unlike Draper's floating beryllium sphere, the Kearfott system's 'dry' tuned-rotor gyros. The spinning rotor was supported on a shaft direct from the motor, with a not rigid support shaft. The design is that the spring effect of the support shaft is exactly balanced by a "negative spring rate" of the rotor.

Though one advantage the adoption of the four gimbal system was the elimination of a problem with three gimbal systems known as "gimbal lock", a catastrophic system failure caused by two of the gimbals becoming parallel.

The Mk 5 guidance system used the PIPA accelerometers of the Poseidon Mk 3 guidance system. For SINS, the Autonetics (division of Rockwell International) SINS, Mk 2Mod7 was chosen, involving two SINS, monitored by a small, solid core ESG (Electrostatically Supported Gyroscopes).

The MGC had a program stored in 200k PROM (Programmable Read-Only Memory), including the guidance equations and steering laws, along with about 48K plated wire RAM for calculations, and parameters read in prior to launch. Plated wire memory is digital computer memory comprising a 5 mil diameter wire of beryllium-copper plated with an isotropic Permalloy, in an array with an unplated wire and

is useful for aerospace applications such as this.

In other words, plated wire memory has a barrier layer between a core of non-magnetic conductive material, and an outer layer of highly magnetic material, such as Permalloy. The barrier layer is constructed of a conductive material which will not diffuse into the outer magnetic layer, such as gold or gold-copper alloy. Plated wire memory also has the advantage over magnetic core memory in being NDRO (non-destructive read out), and can be read without destroying its contents and requiring a rewrite of the data. [U.S.P. #3,786,448 (1973)] [U.S.P. #3,869,355 (1975), classified 5 years] [U.S.P. #3,906,467 (1975)]

Pershing II (December 1984)

The Pershing II is an order of magnitude improvement in accuracy over the Pershing IA. However, the Pershing II utilizes the Pershing IA's first and second stage motors, launcher, and other ground support equipment. The only change to the Pershing II system is the front end [the RV] of the missile.

The Pershing II guidance computer weight is 35 lbs., a size of 12.5" x 13.75" x 7.5", and has a brazed aluminum housing. It is powered by a current of approximately 30 vDC, with a forced air cooling system. It is centered around a parallel, microprogrammed 16 bit microprocessor (CPU), with a 2.4 µs add time. It has a 16k x 16 bits plus parity RAM and ROM memory with a 1.2 µs cycle time. There are 13 priority interrupt levels, pulse inputs from the 3 accelerometers, and serial inputs to a 16 bit serial to parallel converter.

The secondary navigation system (the Sensor Correlator Subsystem), a terrain imaging system, is started as the RV approaches the target, causing the Pershing II computer to generate steering commands to the

RV air vanes (fins) in the final approach to correct errors in the missile's inertial guidance system data. The SCS is a combination radar sensor and automatic area correlator. The SCS consists of the Stabilized Antenna Unit (SAU), Radar Unit (RA), Correlator Unit (CU), and Power Converter Unit (PCU).

The SAU consists of a three-degree-of-freedom gimbaled antenna with associated drive motor and resolver pickoffs. The SAU utilizes a separate altitude measurement antenna and a terrain mapping antenna. The unit is 16" high and 16" in diameter, is mounted forward in the nose of the missile and is covered by the radome.

The Pershing II radome on the tip of the missile consists of a titanium-zirconium-molybdenum (TZM) alloy nose cap subassembly (including an impact fuse and a stealth RF energy absorber), a finely ground slip-cast fused silica (SCFS) ceramic radome body, and a glass-epoxy attachment ring which is bonded to the base of the ceramic shell. The ceramic is spray-coated with a sealer of silicone resin. [AD-A017 242 (1975)]

The Pershing II saw the development of a new and improved automatic gyrocompassing technique. The Pershing II has a linear acceleration of around 40 g's. This resulted in a more accurate azimuth self-alignment. Gyrocompassing is performed on the 4-gimballed platform (azimuth, inner pitch, roll, and outer pitch) IMU before missile launching. This operation establishes the initial alignment of the IMU platform coordinates with respect to a geodetic coordinate system. The geodetic reference adopted is a triad (N, E, and A) where three reference axes are pointed northward, eastward (with the X and Y axes of the horizontal plane), and vertically downward.

The gyrocompassing scheme adopted is referred to as "two-position offset zero-torquing gyrocompassing." [ AD-A067 516 (1978)]

Accurate gyrocompassing can only be achieved if an accurate determination of the east gyro drift can be obtained. Accurate drift determination can be made only for north gyro, a two position scheme can be devised to take this advantage. First, auto-biasing the north gyro to determine north gyro drift. Then the platform is slewed approximately 90° so that the north gyro becomes an east gyro, and the original east gyro becomes a south gyro. Gyrocompassing is performed with the present east gyro, whose drift has been accurately determined, enabling an accurate azimuth alignment. After slewing, the south gyro is in a favorable position for accurate drift determination. The autobiased south gyro will then be ready for in-flight navigation. [AD-A002 696 (1974)]

The all-weather radar at the tip of the missile/RV is activated in the terminal phase of the ballistic path to correlate the returns from an area surrounding the target with a prestored reference map of the target area. Several such correlations are obtained during terminal descent to derive positive corrections for updating the inertial position of the RV. Basic guidance is provided by a combination of inputs from the inertial measurement and radar correlator systems. Control is maintained by both reaction jets and air vanes at the base of the RV. Guidance commands are given by the Pershing airborne computer which derives its inputs from the IMU, rate gyro unit, and the radar correlator system.

The IMU was built by Singer-Kearfott and uses off-the-shelf hardware and designs from the KT-70 family of IMU's, principally from the SRAM missile and A-7 aircraft's IMU's (gimbal system, gyros and accelerometers). There is a forced air cooling system. The IMU weighs 32 lbs., and is 8.5"x10.5"x13" in size.

MX/Peacekeeper ICBM (1986)

The advantages of this ICBM were its highest targeting accuracy in history of any ICBM or SLBM, and its high throw-weight, allowing for the lifting of 10 MIRV high-yield W87 475 kt warheads.

Trident II SLBM (1989)

The Mk 6 INS for Trident II was a much improved model and with superior performance. It has 10 PIGAs and the FLIMBAL, fluid beryllium sphere gyro.

| Solid-fuel ICBM/SLBM | Year First Deployed | ICBM CEP (nm) | SLBM CEP (nm) |
|---|---|---|---|
| Polaris A-1 | 1960 | | 2.0 |
| MM I | 1962 | 1.1 - 1.5 | |
| Polaris A-2 | 1962 | | 2.0 |
| Polaris A-3 | 1964 | | 0.5 |
| MM II | 1966 | 0.43 | |
| MM III | 1970 | 0.25 | |
| Poseidon | 1971 | | 0.25/0.28 |
| MM III Upgrade | 1979 | 0.12/0.10 | |
| Trident I (C4) | 1979 | | 0.12 |
| MX/Peacekeeper | 1986 | 0.06 | |
| Trident II (D5) | 1989 | | 0.08 |

[MacKenzie p. 167, 205, 241 (1991)] [Stumpf, p. 185 (2000)] ["U.S. Strategic Objectives and Force Posture" (1971)] [Spinardi, p. 139 & 146 (1994)]

\* \* \* \*

An issue first raised in the late 1960's by a RAND Corp. engineer named Hyman Schulman, later a Minuteman III guidance system designer, J. Edward Anderson, reported rather inconveniently that ICBM accuracy was overstated.

It was all purely theoretical, being based on computer modeling of data collected from test firings to the "Western Missile Test Range", from launches at Vandenberg AF base in California, 4,300 nm (8,300 km) over the Pacific Ocean, to an RV splash-down near or in the lagoon of the large Kwajalein Atoll, in the Marshall Islands.

But in real use, rather than an east to west trajectory, the ICBM's would travel north-south over the North Pole, which exerted different, untested, and powerful magnetic and gravitational forces on the missiles.

His calculations showed that any error, or discrepancy in the mapping of the gravitational field of the Earth of just three parts in a million (0.0003%) would cause a missile to go off course from its target by 100 yards. [Coates, p. 145 (1985)] [Kaplan, p. 374 – 375 (1983)]

The ICBM's polar trajectory would also be affected by the increased and variable solar and cosmic radiation, which would affect ("bias") its INS guidance system.

There would also have to be continuing exact adjustments to account for the position of the moon at launch, as well as atmospheric re-entry conditions (air density, wind speed, rain, and cloud cover), all of which affect the trajectory.

Kosta Tsipis (1987) enumerated the targeting errors of ICBMs and SLBMs: errors in initial conditions (position error, velocity error, error in vertical alignment), inertial sensing errors during boost (powered flight) phase (accelerometer bias, accelerometer calibration error, initial misalignment of accelerometer axes, misalignment of accelerometer axes due to flight vibrations, initial gyroscope drift, gyroscope drift due to acceleration during flight, gyroscopic drift due to flight vibration), errors in thrust termination, errors due to gravitational anomalies during coasting portion of flight, uncertainty in location of target, unguided reentry effects. The major causes of inaccuracies are velocity error, error in vertical alignment, accelerometer calibration error, gyroscopic drift due to flight acceleration, and errors in thrust termination.

\*   \*   \*   \*

Established in 1972, the Defense Mapping Agency (DMA), was renamed the National Geospatial-Intelligence Agency (NGA), in Arnold, Missouri. In 2003, it was again renamed the National Imagery and Mapping Agency (NIMA). It supports the DoD, and its HQ was in Bethesda, Maryland, but moved to a new HQ in Springfield, Virginia.

NIMA produces geo-spatial intelligence (GEOINT) of the physical features of the Earth's geography. Working with the NRO's satellite capabilities, and the U.S. Geologic Survey [Priest, p. 67-69, 71 (2011)] was responsible for ICBM trajectory variances due to the earth's gravity and magnetic field.

Geodesy, the purview of the Defense Mapping Agency (founded in 1972), is defined as the "branch of applied mathematics which determine by observation and measurement the exact positions of points...[on] the earth's surface, the shape and size of the earth, and the variations of terrestrial gravity."

For example, there's the magnetic North Pole, and then there is the "actual" North Pole. And they're located at different points. The "actual" North Pole is the top point of the axis of the planet Earth's physical spin (one rotation per day!), and the magnetic North Pole, which is where a magnetic compass points to.

The magnetic North Pole's position is determined by the independent spin of the Earth's inner, molten iron core, located miles below the surface. The magnetic North Pole is at 75.5° North latitude, and 100.5° West longitude. The actual North Pole is at 90° North latitude period, a difference of 14.5°! (There is no longitude for the actual North Pole).

The geo-magnetic field intensity over most of the polar region is greater than 0.55 gauss, and is greater than 0.60 gauss at the North Pole (which is 20% greater than the intensity over the northern United States). [AD-A197 814 (1980)]

For another example, the circumference of the earth (the length of the equator) is accepted as being 24,901 miles in diameter. The "geoid" is defined as the contour line or surface along which the gravity potential is the same everywhere. The first real geodetic satellite launched was ANNA-1B in 1962. It used the new SECOR system. Radio signals change phase slowly, the further they travel. A ground station can measure the phase change, and thus the range. A Doppler system was also used for measurements. [AD-A142 764 (1983)].

"Before the imposition of any requirements from the cruise missile project, the DMA possessed the technical capability (equipment and techniques) to produce digital data for terrain following and TERCOM ("terrain contour matching," first patented in 1958 by LTV). The DTED data, used in part for cruise missile terrain following, has been in production since the early 1970's, primarily in support of the Digital Landmass System (DLMS), which provides terrain representations for aircraft simulators. Consequently, little was initially needed in terms of R&D or rate production techniques to support the cruise missile project. In the case of TERCOM, DMA had been producing high quality digital elevation data for developmental flight testing in support of this project since the early 1970s. This basic product was used for TERCOM Aided Inertial Navigation test flights (sponsored by the Navy Cruise Missile Project Office) that were started in March 1973, for contractor's TERCOM test flights, and for the competitive fly-off for the cruise missile guidance system ending in October." [AD-A141 082 (1978)]

"Silicon detectors are related to accelerometers, a critical component of ballistic missile guidance systems." ["Foreign Collection Against the Department of Energy", National Counterintelligence Center et al. (1998)]

In 1988 a Ring Laser Gyro INS was used in the General Dynamics F-16C "Fighting Falcon"

# From Beginning to End

## SAC Bombers, Modern Cruise Missiles, and DSP Satellites

"My life is violent ,
But violence is life.
Peace is a dream,
Reality's a knife."

-- "Colors" (1987)
Ice-T and Afrika Islam

"There is a great deal of [government propaganda] about terrorists.  They are frequently described as mindless, irrational killers.  But terrorism for the most part is not mindless violence....

Terrorism is violence for effect.  Terrorists choreograph violence to achieve maximum publicity.

Terrorism is theater."

-- "Will Terrorists Go Nuclear?" (1975)
Rand Corp. Report P-5541
Brian M. Jenkins

**Strategic Nuclear Heavy Bombers**

"Unfortunately, the B-29 is a splendid plane."

-- Japanese Empress Nagako (1945)
in a letter, shortly after Hiroshima

Boeing B-47 Stratojet (B-47B and B-47E-II Stratojet) (nuclear use:  1951 – 1965)

White underflash areas painted on underside of fuselage for nuclear thermal radiation protection.  It could carry a bomb load of 25,000 lbs., of Mk 5, 6, 15, 18 (B-47B), and Mk 28, 36, 43, and only one Mk 36, one Mk 41, or one B53, carried internally.

Combat Radius:  2,050 nm (B-47E)
116 ft. wingspan, wing area of 1,428 sq.ft., and swept wings
6 General Electric J-47-GE-25 turbojet engines (B-47E)
IFR capability
Aerojet JATO (Jet-Assisted Take Off)

Required forward Air Force bases to be able to hit Soviet targets. The bases were mostly spread along the Soviet periphery, and included Goose Bay, Labrador and the nearby island province of Newfoundland (both in Canada), the U.K, Norway, Greece, Italy, Turkey, Ascension Island (U.K.), Okinawa (Japan), Hawaii, Iceland, Thule (Greenland), Spain, Spanish Morocco, the Atlantic Ocean and Mediterranean Sea islands of the Azores (Portugal), Bermuda (U.K.), Malta, and Cyprus (Greece), Libya, Egypt, Saudi Arabia, and the western Atlantic Ocean island of Puerto Rico (U.S.). A typical Air Force Base was 5 square miles in area. It took the refueling by five KC-97 tanker planes to double the combat radius of a B-47.

An average of three B-47's were always on 15-minute alert, two in Alaska, and one based in Guam. 88% of the strategic targets were located in the western portion of the Soviet Union.

Unfortunately, the forward bases were exceedingly vulnerable to pre-emptive attack by the Soviets, because of their proximity to Soviet territory. It was a concern raised by Albert J. Wohlstetter, a mathematical logician of the RAND Corporation's Economics Division, in his reports, R-266, "Selection and Use of Strategic Air Bases" in April 1954, and in his R-290 study in 1956 that were completely ignored by SAC. [Kaplan, p. 202 – 203 (1983)] [AD-A353 633 (1998)]

The B-47B (first in service in October 1952) was only suitable for carrying atomic weapons, which is what it had been designed for. The B-47E went into service in April 1953, and 1,341 were built. The first B-47 capable of carrying thermonuclear weapons (H-bombs) became available in January 1955. By the end of April 1956, it began handling the new thermonuclear bombs.

Around 1960, B-47's were also based in eleven U.K. SAC-controlled Air Force bases, and three bases in Spanish Morocco. B-47 bombers were currently (1960) deployed at 10 air bases on foreign soil: four in the U.K., 3 in Morocco, and 3 in Spain. [WSEG No. 50 (1960)]

In mid-1957 it was proposed to allow the carrying of thermonuclear bombs (rather than just the Mk 6 fission bomb) by bombers when stationed in non-U.S. [foreign] bases. ["History of the Custody and Deployment of Nuclear Weapons," p. 49 (1978)] But it could carry less atomic bombs than the upcoming B-52, as well as because of weight-carrying limitations it was less suitable to deliver hydrogen bombs. [AD-A209 273 (1988)]

Boeing B-52 Stratofortress Strategic Heavy Bomber (1955 – 1991)

Dimensions:  48' high x 152' long; 185' 9" wingspan, swept wings, which keeps it more stable at its top speed.
Speed:  Mach 0.84
Propulsion:  8 Pratt & Whitney TF33-P-3/103 low-bypass turbofan engines; there are 4 engine pods, each containing  a pair of engines; each engine providing more than 17,000 lbs. of thrust for the H Model (i.e., the B-52H).
Range:  Combat radius of over 4,100 miles, while carrying a 10,000 bomb load.
Other Capabilites:  IFR (Inflight Refueling), and ECM (Electronic Counter-Measures)

From 1955 to the early 1960's, the only B-52 Strategic Air Command (SAC) base outside of the continental U.S. soil (plus Alaska and Hawaii in the mid-Pacific),  was in American owned Puerto Rico, an island in the Caribbean Sea, south of Florida.

There were nine in-flight refueling tanker airplane facilities for use during war:  six in Canada, and one each in Thule (on the coast of western Greenland, owned by Denmark, for northern flights across the North Pole region to attack central and eastern Russia (such as Vladivostok), and Red China, and

Bermuda (a U.K. mid-Atlantic possession), and the Azores (owned by Spain), all three across the northern Atlantic Ocean.

Bermuda and the Azores were for B-52 attacks against Moscow — the capital, command center, and principle target of the U.S. — as well as missile silos in the Ural mountain range, and the other major cities of western Russia, as well as the communist Warsaw Pact alliance nations in eastern Europe (such as Poland and East Germany).

In the case of a general (i.e, total) nuclear war, there were also available four air bases in the U.K., one in Turkey (for attacks on eastern Russia and its allies), two in Japan, and one each in the Japanese owned island of Okinawa (1960), all three to support attacks on western Russia and Red China.

The B-52G & B-52H burned 30 tons of jet fuel for its magic 10,000 mile maximum range (for a total forward and backwards distance of 20,000 miles).  It burned 30 tons of fuel per hour.  (2 x 10,000 miles total flight distance) x (30 tons/hour) of fuel / (600 m.p.h. speed) = 1,000 tons of fuel consumed after sometimes repeated in-flight refueling.  [However, the return trip after the bomb(s) had been dropped on Russia and China would probably involve only relanding empty in Japan, Okinawa, Hawaii, or Alaska, for instance.

The B-52's first year was 1955.  A B-52B in May 1956 was the first plane to drop an H-bomb, the TX-15-X1 (later redesignated the Mk39), weighing 7,000 lbs. and with a yield of 3.8 MT.  The B-52C was introduced in June 1956, but only a small number (35 bombers) were manufactured.

The B-52D with a payload capacity of 13,500 lbs., could carry the newly developed H-bombs, and it was introduced into SAC bomber fleet in December 1956.

A few B-52E's began reaching SAC in December 1957.  They were not that much different from the B-52D model.

The next model, the B-52F could carry a payload weighing 38,000 lbs.  It began reaching SAC service in June of 1958.

The B-52D/E/F/G & H could carry a load of either multiple Mk 28 or Mk 57 bombs, or only one of the heavier Mk 41 or Mk 53 nuclear bombs.  The B-52G that crashed near Goldsboro, North Carolina in 1961 carried two Mk 39 bombs.  Most of the B-52's in the 1960's carried four Mk 28 bombs.

The first major changes to the B-52, were with the B-52G & H.  The B–52G and B–52H, with new aluminum wings, were lighter than preceeding B-52 models, and had larger fuel loads.  The B-52G (first deployed in 1959) & H could carry a bomb load of 25,500 lbs.  The B-52G has a "Hi-Lo-Hi" altitude (incoming/bombing mission/outging escape) bombing range of 3,200 nm, and also carried 2 defensive Hound Dog Mk 28-armed missiles externally suspended on both wings, until they began to be replaced/upgraded to the much lighter and superior SRAM, W-69-equipped missile in 1972, with up to 20 SRAM's per B-52.

The B-52H had a increased combat radius range of 3,550 nm.  It featured Pratt & Whitney, 17,000 lb. thrust, TF-33-P-3 turbofan engines (without the previously equipped water injection system), and was introduced to service in May 1961.  [AD-A209 273 (1988)]

The radar cross section of the enormous B-52H is only 100 m$^2$, compared to 7 m$^2$ for the FB-111A.  [AD-A532 565 (2010)]

General Dynamics (was Convair) B-58A Hustler (1960 – 1969)

Dimensions:  31' 5" high x 96' 9" long
Wingspan:  56' 10"; Delta-wing aircraft
Propulsion:  Four GE J-79-GE-5A or -5B afterburning turbojet engines
Range:  4,450 miles
Speed:  Mach 2/1,321 mph
Thrust:  15,600 lbs. per engine
Use:  A medium bomber, capable of in-flight refueling.  116 built.  From 1960 to phase-out beginning in 1969, to be replaced by the FB-111.

First U.S. supersonic strategic bomber.  One nuclear load was 1 B-53 and 4 Mk 43 nuclear weapons. [Rhinehammer (1964)]  About 100 were built, but the B-58 proved unsatisfactory, and was retired early.

*   *   *   *

Numbers of Nuclear-Capable Bombers [AD-A602 158 (2012)]:

| | B-29 | B-50 | B-36 | B-47 | B-52 | B-58 | Nuclear Payload |
|---|---|---|---|---|---|---|---|
| 1947: | 319 | | | | | | Mk4 |
| 1948: | 486 | 35 | 35 | | | | Mk4 |
| 1949: | 390 | 99 | 36 | | | | Mk4 |
| 1950: | 286 | 196 | 38 | | | | Mk4 |
| 1951: | 340 | 219 | 98 | 12 | | | Mk6 |
| 1952: | 417 | 224 | 154 | 62 | | | Mk6 |
| 1953: | 110 | 138 | 185 | 329 | | | Mk6 |
| 1954: | | 78 | 209 | 795 | | | Mk15/Mk18 |
| 1955: | | | 205 | 1,086 | 18 | | B-47:  Mk15/39/Mk17/Mk24 |
| 1956: | | | 247 | 1,306 | 97 | | B-47 & B-52:  Mk15/39/Mk17/Mk24 |
| 1957: | | | 127 | 1,285 | 243 | | B-47 & B-52:  Mk15/39 |
| 1958: | | | 22 | 1,367 | 380 | | B-47 & B-52:  Mk28/Mk36 |
| 1959: | | | | 1,366 | 488 | | B-52:  Mk28/Mk36 |
| 1960: | | | | 1,178 | 538 | 19 | B-52:  Mk28/Mk36 |

*   *   *   *

The final conquest of the Marianas in 1944 was the culmination of the Pacific War against the Japanese Empire by the U.S. Tinian Island in the Marianas, 1,500 miles from Japan, became the base for long-range bomber attacks on the home islands of Japan.

The B-29 Super-fortress that bombed Hiroshima from Tinian Island, was just over 141 feet wingtip-to-wingtip, and its basic weight was 74,500 lbs., but fully armed it could weigh  135,000 lbs.  It could deliver a ten ton payload 1,500 miles.  Its four 2,200-horse-power engines drove propellers 16.5 feet in diameter.

It was replaced in 1948 (it first flew in 1942) by the Convair B-36 Peacemaker, a longer range, nuclear-capable heavy bomber, that was still just a stop-gap measure (awaiting the introduction of the B-47 and B-52).  A fleet of 342 B-36's was the peak number in 1954.  The last B-36 was retired in 1959.  [Goldberg, part 2, p. 877 (1981)]  It used six piston motors for its propellers, rather than the new jet engines (first fielded by the Germans in late WW II, the ME-262, with two jet engines under the straight wings) because jet engines guzzled too much fuel for the range needed for intercontinental bombing.  It was capable of a flight of 6,800 – 8,175 miles, with aerial refueling required.  [Weitze (1999)]

Jet aircraft goes back to July 1926, when a young British engineer name A.A. Griffith published a paper on jet turbines.  The idea was followed up by Frank Whittle, who in January 1930 took out a patent on the first jet engine.

In 1936 Hans Joachim Pabst von Ohain, a gifted young engineer, took out a patent for the use of the exhaust thrust from a gas turbine as a means of propulsion. Ohain presented his ideas to Heinkel, and the first engine was successfully demonstrated in 1937. On August 1940, the idea was demonstrated in public with the Heinikel He-178 V1, but the hydrogen fuel burned through the engine. His HeS-3 burned diesel fuel, Ohain's third design.

Simultaneously, along with the British and Germans, the Russians were also at work on jet propulsion. Arkhip Lyulka from Kiev Oblast in the Ukraine first work was on turbofans as superchargers of piston engines. But between 1939 and 1941, he worked on what was to become the first double-jet turbofan engine in the world, which he patented in April 1941. The Russians used German designs copied from Junkers and BMW jet engines brought to Russia from Germany after the war. The U.K. govt. also exported 40 Roll-Royce Nene turbojet engines, which the Russians pirated to make the MiG-15, used later in the Korean War, which proved to be superior to anything in the West. The MiG's were built by a company established in Dec. 1939 by Artem Mikoyan, a young aviation designer from Sanahin, Armenia.

Messerschmitt ME-262 _Schwalbe_ (Swallow) or _Sturmvogel_ (Stormbird) both came out at almost the same time: July 1944. The ME-262 was vastly superior, with a speed of 560 mph (900 kmph) vs. 410 mph (660 kmph) for the Gloster Meteor. The ME-262 had a 5:1 shoot down ratio over Allied fighters.

German design went on to influence the construction of the Boeing B-47, as von Ohain was brought to the U.S. in 1947 under Operation Paperclip, to join the staff of Wright-Patterson AFB, of which he became the director of the Aeronautical Research Laboratory in 1956.

Whittle later wrote a book entitled "Gas Turbine Aero-Thermodynamics" which appeared in 1981.

Eugene Sanger, born in Pressnitz, now in the Czech Republic, an aeronautical engineer published his rejected university thesis as "Rocket Flight Technology" in 1933. Worked with Irene Bredt of Vienna, a brilliant young mathematician, whom he later married. Their work and calculations contributed significantly to the U.S. North American X-15 rocket-plane.

In the post-war race to develop the jet engine equipped long range bombers were three contenders: General Electric, Westinghouse, and the English company, Pratt & Whitney. Westinghouse had lengthy experience with steam turbine technology for power generation, and which closely resembled the components used in turbo-jets.

But Pratt & Whitney, raced forward from behind, with the new concept of a "twin-spool" turbo-jet engine, which had two spinning turbines, one in back of the turn the front turbine, to boost performance and efficiency by compressing its air intake to twelve atmosphere pressure, resulting in more miles per gallon fuel. This allowed the new engine, called the J-57, to power heavy bombers to intercontinental range.

In 1948, SAC, the Strategic Air Command was formed. It was placed at Offutt Air Force Base in Nebraska. In 1992, SAC was renamed the U.S. STRATCOM, the U.S. Strategic Command, which is located at Bellevue, a small town just south of Omaha, Nebraska.

"As of mid-1950, the nuclear strike force consisted of more than 200 B-29s, B-50s, and B-36s, plus a handful of Navy aircraft. By 1953, it included more than 1,000 Air Force and Navy aircraft." [Goldberg (1981)]

The Boeing B-47 Stratojet, the first jet-engine powered bomber, followed the B-36, and became the backbone/workhorse of SAC during most of the 1950's, first becoming operational in 1951. In early 1951 it was in-flight refueled successfully by a Boeing KC-97A Stratofreighter tanker aircraft, greatly increasing its range.

From President Truman, and continuing with President Eisenhower, from 1948 to 1953, the number of long range bombers increased dramatically from 30 to 1,000. The B-47 with its new jet engines, also had a new aerodynamic shape with a new innovation: swept-back (angled) wings which reduced drag and allowed for higher speed to take advantage of the power of its jet engines. (The swept-wing design had been recovered from Nazi research on the issue.) And it was also the first nuclear bomber that was made intercontinental by in-flight refueling by a new fleet of KC-97 tanker aircraft.

The SAC fleet of B-47's, which began overseas alert operations in October 1957 [Narducci (1988)], peaked in numbers at 1,800 in 1957, but was by then already about to be eclipsed by the B-52, which became operational in 1955, with total SAC forces peaking in 1958. The B-52 had eight Pratt & Whitney jet engines and could carry a thirty ton load of bombs, versus the B-47's six jet engines and ten ton bomb load.

The last B-52 model, the B-52H, entered service in 1961, by which time the turbo-jet engines had been replaced by the considerably more fuel efficient (and thus longer range) high bypass turbofan engines (later also adopted by the modern Cruise Missile), with a revolutionary high bypass of a ratio of 8:1, resulting in extremely low fuel consumption.

The Air Force Propulsion Laboratory, working with and in parallel with GE, discovered that the key to getting a high-bypass fan working was to increase the turbine inlet temperature, which required improved turbines, achieved by better blade cooling. [AD-A528 970 (2005)]

As well as one B-53 (8.9 MT) bomb, B-52's have carried verified/declassified bomb loads of 4 Mk 28's (1966 Palomares, Spain and 1968 Thule, Greenland incidents), and 2 Mk 39's (1961 Goldsboro, North Carolina incident). The incidents were plane crashes.

With the entry of the B-52H in 1961, the B-47 began to be retired, and was gone by 1966. At SACs peak strength, in 1958, they had a fleet of more than 2,000 B-47 and B-52 bombers.

At the end of 1961, SAC had 800 B-47's and 550 B-52's, down from a peak of 1,370 B-47's at the end of 1958. The B-47 was due to be phased out completely in 1966, and the B-52 production stopped in 1962, with peak numbers of 630 B-52." [Goldberg (1981)] At the end of 1963, SAC had 1,300 bombers, of which half were on 15 minute ground alert at 53 bases. By 1967 SAC was down to 697 bombers, [Wainstain (1975)] and by 1973, the B-52H strategic bomber fleet was down to 397 planes. [JCAE, Pt. 1, p. 7 (1973)]

Though, by 1965, eclipsed by the Navy's Polaris SLBM in 1960 and the Air Force's new Minuteman I in 1962, there were just 500 B-52's remaining in the SAC fleet. [Addington, p. 263, 272 (1984)]

From 1953 - 1968, when the rest of the nuclear Triad (ICBMs/SLBMs/Navy carriers), and the 1966 Palomares & 1968 Thule nuclear accidents led to the cancellation of Operation Chrome Dome (which ran from 1960 – 1968) with planes always in the air 24 hours a day on full nuclear alert. Operation Chrome Dome were 24-hour missions with two in-flight refuelings [Nijboer (2003)] – each refueling over 100,000 lbs. of fuel in about 20 minutes [Miller, p. 151 (2010)] -- B-52's on alert typically carried four Mk 28 thermonuclear bombs as payload.

And the introduction of the 700 lb. B61 for B-52's and carrier-based bombers and IFR (Inflight Refueling) of the B-52 made politically-sensitive foreign bases expensive, and superfluous.

* * * *

The B-52 was the longest-lived warplane ever produced, and is still in service. Pilot's SAC flying checklist was an incredible 42 pages long. The vulnerable hydraulic-powered flight control system of the B-47 was replaced by manual force: the control stick for the wing ailerons required 37 lbs. of force from the pilot.

The B-52E, first fielded in late 1957 carried externally two of the new AGM-28 Hound Dog nuclear "stand-off" missiles, as well as its load of Mk 28 nuclear bombs. The latest model of B-52, the B-52H was first launched in 1961.

A typical B-52 bomb load in 1973 was four B61 bombs and eight SRAM missiles in the internal rotary rack. [JCAE, Part 1, p. 7 (1973)] Its maximum load was upped to twenty B61's, from twelve with the B61 and two Hound Dog missiles. [Gervasi, p. 86 (1984)]

The U.S. had also thousands of tactical aircraft based in Western Europe to deliver smaller bombs onto Russian or East European targets, and over 400 carrier-based warplanes that could hit Soviet targets from its Eastern end..

But the B-52 was the post-card image of the nuclear Triad, known to every American, as the great American nuclear shield. "Peace is our Profession" was their laughable motto, though their business was actually mass murder on a grand scale at a moment's notice.

"The material prepared...specially for bomber targeting...includes flight plans for various sorties, lists of Short-Range Attack Missile (SRAM) check-points, instructions for electronic countermeasure (ECM) operations, and target photo information. The bomber flight plans contain detailed instructions for each sortie beginning with the departure base and ending with the post-strike recovery base, with less specific data concerning return home flight from the post-strike recovery base to the continental United States. The SRAM Checkpoint Master List contains all the checkpoint locations for updating navigation system. The instructions for operation ECM specify the equipment-time-location factors for use of ECM. The target photo material includes simulated radar photos for both targets and checkpoints." [DNA-6147T (1982)]

But the B-52 was the obsolete program that just would not die. Inferior to the hide-able SLBM, and 30 minute flight time of unstoppable ICBM in 1960 and 1962, respectively. IFR replaced the pressure from foreign countries nervous about hosting them. Then Hound Dog, then SRAM, then the ALCM stand-off missiles enabled them to destroy Soviet border air defense installations and penetrate to their targets.

ECM equipment was added, day-light attacks were replaced by nighttime assaults, and high-level flight was replaced by low altitude flight to evade air defenses.

Finally B-52-carried Air-Launched Cruise Missiles (ALCM's) allowed the B-52 to not even have to penetrate the Soviet border. Eclipsed long before by sub-launched Cruise Missiles (SLCM's), SAC was finally abolished by President H.W. Bush in 1991.

\* \* \* \*

On the other hand, the Soviets had the best border air defense network in the world. In late 1989, they had about 10,000 radar installations, about 8,000 SAM sites, and about 3,000 fighter/ interceptor warplanes and AWACS control airplanes. [Scott (1991)]

Starting in 1966, with the ZSU-66 radar-controlled 23 mm full auto cannon, with a range of 3 km, then SA-6, the first SAM, with a range of 24 km. Then the 1974 SA-8, with a range of 15 km, which was widely deployed, followed by the SA-12A with a range of 80 km.

And finally the 1989 2S6, with a radar-controlled 30 mm cannon and four IR-guided SA-19 SAM's with a range of 8 km, and an independently-located radar, to avoid perimeter anti-radar missiles such as the B-52-carried nuclear SRAM. [Dunnigan, p. 120, 123-125 (1990)]

**Fighter-Bombers**

FB-111 medium bombers and the F-4 and F16 fighters, which are based in Europe, can deliver nuclear weapons to the western USSR [Panofsky, p. 60 (1979)], or from aircraft carriers (first achieved in 1955 with the U.S.S Forestal. Each carrier at the time (ca. 1960) carried about 36 tactical nuclear aircraft and

100 nuclear weapons, with an ammunition ship or two somewhere close by with another 100 nuclear weapons available for the carrier in each ship.  [Miller, p. 69 (2010)]

North American Aviation F-100 Super Sabre (tactical nuclear use:  1955 – 1970)

Bomb Load:  7,000 lbs.

F-100A & F-100C:  one Mk 7.
F-100D:  one Mk 28, Mk 43, Mk 57, or B61.
F-100F:  one Mk 7, or one Mk 28, Mk 43, Mk 57,  or B61; combat radius:  450 nm

The world's first production supersonic airplane.

Republic F-84G Thunderjet (1955 – 1972)

G-model first to be able to carry nuclear bombs
High sub-sonic speed; swept wing
Wright J65-A-29 turbojet engine.
Wingspan:  38' 1"
Weight (empty):  11,095 lbs.; Max. Gross Weight (including weapons):  23,525 lbs.
Could only carry 1 Mk 28 nuclear weapon, or in 1966 two B61's.
Combat radius:  810 miles (1,304 km)
Bomb load:  1,800 lbs.

Became the first USAF jet fighter capable of carrying a tactical nuclear weapon.

Northrop F-89J Scorpion (Jan. 1957 - 1968)

J-model first to be able to carry nuclear rockets.
Subsonic Fighter-Interceptor (all-weather; day or night).
Engine:  2J35-A-35
Wingspan:  59' 8"
Takeoff Weight:  45,575 lbs.
Combat Radius:  435 nm

Was the Air Force's first nuclear-armed interceptor; carried two MB-1 Genie (W-25) air-to-air nuclear rockets.

McDonnell F-101 Voodoo (1957- 1971)

F-101A & F-101C were nuclear-capable supersonic tactical fighter-bombers
One Mk 7, Mk 28, or Mk 43.

Convair F-102A Delta Dagger (June 1958 - 1972 )

Supersonic, all-weather fighter-interceptor.
Combat radius: 566 nm
Equipped with nuclear Falcon air-to-air missile (named GAR-11, then re-designated the AIM-26A).

McDonnell Douglas F-4 Phantom (and Phantom II) (1958 – 1991)

5 ton payload capacity; Can carry three Mk 57's, or later, three B61's.
Maximum Speed:  Mach 2
Maximum Range:  1718 nm (3184 km)
5,200 built.

Republic F-105 Thunderchief (August 1958 – retirement 1975 to February 1984)

Supersonic, fighter-bomber.  Originally designed to deliver only nuclear payloads in a tactical mission. Could carry four Mk 57 nuclear weapons or two Mk 28, Mk 43, or B61's carried externally.  The F-105D could carry over 12,000 lbs. of ordnance.

Power Plant:  One P&W J75-P-19W engine
Thrust:  23,500 pounds
Dimensions:  19' 8" high x 63' 1" long with a wing span of 34' 11"
Maximum Speed:  1,375 mph
Range:  2,200 miles
Number:  833 built

General Dynamics FB-111 (F-111F Aardvark) (1969 – 1991)

Twin jet, supersonic, swing-wing medium bomber; replaced the B-58.  A strategic and tactical bomber version of the F-111.

2 B43 standard, then B61's after 1966; then SRAMs after 1972 (two internally and four on external pylons. [LA-11401, p. 21 (1991)] [JCAE (1973)]

Maximum load:  31,500 lbs. (14,228 kg).  Can carry six Mk 57s
Power Plant:  Two Pratt & Whitney TF-30-P103 turbofan engines
Thrust:  18,500 lbs. per engine
Dimensions:  17' high x 73' 6" long
Wingspan:  32' swept
Maximum Speed:  1,452 mph
Range:  3,000 nm; 700 nm, without refueling; 2,540 nm (2,925 miles; 4,700 km) with.
Payload:  9,000 lbs.
[Goldberg, part 2, p. 845 (1981)]

Used from at least Vietnam (1965) to Gulf War I

General Dynamics F-16

First built in late 1970's. Nuclear capable, weighs less than one half the F-4 Phantom, accelerates twice as fast, and requires one-third the maintenance hours as the F-4, which the F-16 replaced. [AD-A065 881 (1978)]

Lockheed F-117A Nighthawk

The first stealth fighter
Subsonic
5,000 lb. (2,300 kg) maximum bomb load; can carry 4 bombs internally
nuclear strike role
Combat radius: 465 nm (535 miles; 856 km)

Northrup Grumman B-2A Spirit (Stealth Bomber)

Flying wing, with no tail.

Length: 69 ft. (21 m); Wingspan: 172 ft. (52.4 m); 10 ft. thick, at its thickest point.
4 GE turbofan engines.
Range with an IFR: 10,000 nm (11,500 miles; 18,500 km)
Range unrefueled with 8 SRAM and 8 B83 bombs: 4,400 nm (5,075 miles; 8,170 km)

All carbon composite leading edge and skin paneling; triangular construction; large double exhaust vents, large area, bottom-lined to radiate upwards.
LPI (Low probability of intercept) J-band radar unit (probably lead spread spectrum /frequency-hopping)

Designed with a strategic nuclear bombing mission in mind. 2 internal rotary launchers for 16 ACM's or SRAM's; can carry 16 B83 or B61 bombs
Bomb load capacity: 44,800 lbs.
Cost: a whopping $500 million each. [Scott (1991)]

Lockheed F-22A Stealth Fighter-Interceptor

In Service: 1997 to 2013 at least

**Naval Aircraft and Naval Nuclear Weapons**

Of interest to nuclear weapons in the U.S. Navy are, in order of decreasing size: the aircraft carrier, the battleship, and the cruiser (heavy). 19 aircraft carriers have been built since 1945: 7 Essex class and 4 Forrestal class in the mid-1950s, and eight nuclear-powered Nimitz class carriers, starting in 1968. [Gardiner, p. 37 (1993)] In January 1955, there were five naval vessels in the Pacific, and three naval vessels in the Atlantic Ocean that carried nuclear weapons. ["History of the Custody and Deployment of Nuclear Weapons" (1978)]

## Nuclear Capable Aircraft Carriers

| Class | Midway | Forrestal | Kitty Hawk | Enterprise | Nimitz |
|---|---|---|---|---|---|
| IOC | 1945 | 1955 | 1961 | 1961 | 1975 |
| Number | 2 | 4 | 4 | 1 | 8 |

Nuclear Weapons: All carriers had at varying times the Mk 7 and Mk 12, Mk 28, B-43 (>1961), B-57 (>1963), and B61 (>1966), with the Terrier missile for Kitty Hawk. [NWDB-I, p. 251 (1984)]

As of 1996, the Navy had built over 104 nuclear-powered surface ships. Of these, eight aircraft carriers, and four guided missile cruisers were still in operation.

With the introduction of jet aircraft at the end of World War Two in 1945, three innovations were necessary for their use by aircraft carriers: the steam catapult was necessary, the angled deck, and the mirror landing aid were invented by the U.K., and adopted to their fullest by the U.S. By 1952 on the U.S. Navy carrier, Midway had undergone tests, and the fleet carrier Antietam was equipped with a complete eight degree angled deck, proving its value.

By 1954, the Hancock became the first U.S. carrier to be equipped with steam catapults, which were a necessity for jet aircraft takeoff.

By 1955, the U.S. had installed its first mirror landing system with a system of green lights provided the pilot with a view of his landing position. By the 1960's, the mirror system had been upgraded and perfected, allowing the pilot to see and correct his landing angle and position.

The first nuclear carrier, the USS Enterprise was commissioned in November 1961 at a cost of $330 million. It had an eight reactor power-plant. This was soon superseded by the long-lived Nimitz class design using a more efficient 2 reactor power-plant, eight carriers of which have been built starting from commissioning in 1975 of the first one, and 2003, the last one.

While much is made of their ASW helicopter platform role, the use of naval warplanes in strategic offensive operations should not be doubted. Their offensive nuclear use came into use with the introduction the 18" diameter Mk 43 in 1961 - 1965, and the Mk 57, 1963 - 1967, but really came into its own with the 1966 introduction of the 13.5" diameter, 700 lb. Mk 61, a device which did not use 93% enriched uranium, and had a yield of "several hundred" kilotons. [JCAE (1973)].

Thus the role of U.S. naval surface ships is two-fold. It consists of the aircraft carrier flotilla, a movable nuclear platform for ASW use, or strategic offensive use as a platform for naval nuclear-armed aircraft.

The ASW role is assumed by many other ships in the flotilla besides the main battle carrier. Many of these are equipped with the ASRCOC (Anti-submarine rocket), in use since 1961 until the end of the Cold War, and using the 14" diameter W-44 low-kiloton fission weapon.

There was the TALOS RAM-jet powered missile, equipped with the 22" dia. W-30, boosted fission device to neutralize air attacks on the carrier group, as well as coastal targets in a ground-to-ground role. Also in the strategic role was the 8" (203 mm with 16 km range [LA-4350 (1969)]; 18 mile/29 km range; CEP: 800 ft.) W-33 and 6" W-48 artillery shells, of which cruisers were equipped. In 1981, the W-79 ERW 8" artillery shell was added to the list.

The General Dynamics FB-111A swing-wing supersonic fighter-bomber was evaluated in 1970 for use by SAC. It could carry six B61's or six SRAMs (compared to 24 nuclear weapons that can be carried by the B-52G & H), and had an operational range of 4,700 miles, lengthened, of course, if there was inflight refueling.


## Naval (Aircraft Carrier-Based) Tactical Nuclear Attack Bombers

[Goldberg (1981)] [Bowen, Vol.4, part 2, p. 524 (1959)]

Lockheed P2V Neptune (1945 – 1965; modified to become nuclear-capable January 1950): the P2V-5 (the major production version begun in 1950) could carry a Mk 7 bomb load (of 8,000 lbs./3,600 kg); range: 3,500 nm.

North American AJ-1 & AJ-2 Savage: (1949 – 1960); attack bomber; 143 in total built; production completed in 1954; first carrier-based aircraft to be nuclear-armed. Could carry a Mk 6, Mk 13, Mk 18, and anything lighter (like the Mk 5 & 7).

Douglas AD-1/A-1 Skyraider: (December 1946 – 1963; 1953 became nuclear-capable; production terminated 1957); Douglas A2D the follow-on for AJ-2 Savage; 3,155 manufactured. Propeller-driven. Combat radius 1,500 nm; bomb load 10,500 lbs. Tailored to carry a Mk 7 atomic bomb.

Douglas A3D/A-3 Skywarrior (1956 – 1966); twin-engine, swept wing, jet attack bomber; 282 in total built. Replacement/upgrade from AJ-2 Savage bomber. First all jet bomber to be deployed by the Navy. Meant to be a high altitude bomber. Designed to carry the Mk 15/39 thermonuclear bombs. Combat radius with three Mk 27 bombs and upper bomb bay auxiliary fuel tank: 1,410 nm (1,622 miles; 2,610 km). Can safely deliver a ground-burst weapon of greater than 12 MT. [AD339 910 (1956)]

Douglas (later McDonnell Douglas) A4D Skyhawk (1958 – 1976; by late 1960 one thousand had been delivered to the Navy). Bomb load: 3,000 lbs. Could carry the Mk 28, Mk 43, Mk 57, and B61.

North American A3J/A-5 Vigilante (July 1960 – 1968). Range: 2,300 nm. Maximum speed: > Mach 2 at 40,000 feet. Major parts made of titanium. Only 59 built.


## Cruise Missiles: Death on Your Door-Step

The definition of a cruise missile is an unmanned, self-propelled, air-breathing guided vehicle that sustains its flight through aerodynamic lift (via its wings) through most of it flight.

The modern Cruise Missile (CM) is a form of Precision-Guided Munition (PGM). It is not actually a missile, but an unmanned, radar-guided jet, which can carry a nuclear weapon. It flies at a very low level, to avoid air defenses. It is highly accurate, with a very low CEP, its TERCOM radar guidance system exceeding even the most modern ICBM's (the MX/Peacemaker) accuracy. Its radar guidance system evolved from an extrapolation of the radar area correlation (following a previously radar-mapped geographic route) guidance system developed for the Pershing II strategic IRBM in the 1980's.

The original advantages of the early CMs was that they were a one-way, unmanned missile, with no in-flight fueling fleet required, and each Snark, for instance cost 5% of a B-52.

But the CM was a weapon whose time had not yet come. Through the entire 1950's, development efforts of a multitude of CMs came to naught, eclipsed by the Atlas ICBM in 1959, and then in 1960, the infinitely more versatile Polaris SLBM. The only quasi-CM that achieved any length of deployment was the Naval TALOS W surface-to-air missile, which uses the W-30, a 22" diameter boosted atomic device.

The CM would have to wait for new technology, and a more auspicious time for introduction.

That came in 1969, when the Williams WR-19 turbofan engine was developed for another project. The WR-19 became the father of a family of engines that have since been developed for modern cruise missiles. It was a twin-spool bypass fanjet, 24" long and 12" in diameter, weighing about 65 lbs., and producing a thrust of 430 lbs. Most significantly, it was one-tenth the size of the next larger engine. The Williams Research Corporation, of Walled Lake, Michigan, was the major developer of small engines, and was founded in 1954 by ex-Chrysler employee Samuel Williams.

The temperature the engines could tolerate was 1,750 °F in 1970 using Inco 100 alloy for first stage turbine blades. [ADC 021 800 (1980)]

The initial schema for the system was a topological mapping system (U.S.P. #4,359,732 (1982); filed in 1963, classified 19 years).

The second invention was synthetic aperture radar (SAR), which works without requiring a moving, rotating antenna, with a stationary set of copper panels instead. It is ideal as a ground mapping radar because of its ability to resolve very small radar reflective areas [Chrzanowski (1990)], providing photographic-like imagery at long range. SAR provides day/night, all-weather, long range surveillance of desired areas.

An airborne SAR uses the vehicle's own movement to simulate a large radar aperture. It does require significant computing power to process the imagery collected. [AD-A309 937 (1996)]

The ALCM was linked to the cancellation of B-1 bomber by Jimmy Carter (and its predecessor the B-70, canceled by the Kennedy/McNamara admin), which was revived by Carter's successor, Reagan.

\*　　\*　　\*　　\*

But first the beginnings.

A Frenchman, Victor de Karavodine was given French Patent #374,124 for a pulse-jet engine in 1907. In 1919, René Lorin, a French artillery officer, published a paper on using pilotless aircraft driven by a pulse-jet engine to bombard faraway cities.

The V-1 was first conceived in Germany in 1935 and was first launched in Dec. 1942. 9,521 V-1's landed in England, before the final launch site was occupied by the Allies in October 1944. Then Antwerp was attacked with 2,448 missiles, until March 1945 when the last launch site was overrun. Almost 23,000 people killed in the attacks, almost all civilians. Roughly 30,000 V-1's were manufactured, each taking 350 man-hrs., and costing 4% of the price of the V-2 missile. It was cheap, economic to fuel, easy to produce in vast numbers. But they needed a launch ramp, and were easy to shoot down, unlike the V-2.

A Munich fluid dynamics engineer, Paul Schmidt began working on a pulse-jet engine, starting in 1927 - 1928. He was trying to perfect de Karavodine's design, in which a fuel-air mixture was forced into a pipe with spring-loaded valves, and then sparkplug ignited. He had been influenced by a Belgian inventor, George Marconnet, who patented his own jet propulsion work in 1910. Schmidt got his design working by 1939 - 1940.

His pulse jet used an oscillating burning process for gasoline vapor. Once moving forward at a sufficient speed, an array of spring-loaded steel plate flaps/shutters, like venetian blinds, at the front of a length of tubular steel jet pipe, closed, would open by the air pressure, allowing air to enter. Simultaneously, the flaps opened a valve, and a fine mist of gasoline was sprayed from nine pipes, would mix with the air, and then ignited by the sparkplug, exploding. The explosion would slam the front shutters closed and be expelled as a pulse of hot exhaust thrust through the rear end of the tube. The pressure of the air then blew open the vanes, and a new charge of fresh air entered the jet tube. This occurred 47 times per second. [De Maeseneer (2001)] [Ford (2011)] The fuel was fed to the motor by two spherical compressed air tanks at 150 atm. pressure.

One of the crudest, simplest, cheapest weapons ever developed, it weighed 4,750 lbs. (2,150 kg), 27' 3.75" (8.32 m) long, and 17.5' (5.37 m) from wing tip to wing tip, and 4' 8" (1.42 m) tall. Its warhead was 1,870 lbs. (850 kg) of amatol-39 (an ammonium nitrate/TNT mixture), with a max. range of 150 m (250 km) and a flying speed of 400 mph (640 kmph) at an altitude of between 2,000 - 3,000 ft. (600 - 900 m), for access to sufficient oxygen in the atmosphere.

The result was the Nazi V-1 (for Vengeance-1) "buzz bomb", the world's first Cruise Missile (1944). There was one draw-back to the V-1: the primitive RAM jet needed a speed of over 190 mph to work and sustain itself. So, in other words, it couldn't take off on its own. To overcome this problem, the launching ramp, essentially a 150 foot long slotted tube (made from 8 sections), with both ends of the tube open, and the tube supported by steel on a concrete platform at a suitable launch angle ( < 45°).

To launch the V-1, air tanks blew air into the front of the engine, the sparkplug fired, so that the engine started, and ran at full power. After about 7 seconds of this warm-up, potassium permanganate (Z-Stoff) and conc. hydrogen peroxide (T-Stoff) were injected under pressure into the combustion chamber, their reaction together producing a massive volume of super-heated steam. The V-1 was as a result launched at a speed of 250 mph. [Hogg (1999)]

The V-1 was originally fueled by hydrazine hydrate and methanol originally; referred to as C-Stoff (57% methanol, 30% hydrazine, and 13% water), burning in oxygen provided by T-Stoff (80% $H_2O_2$).

Range: 260 km
Top speed: 400 mph (645 kmph)
Fuel:   150 Imperial gallons of low-grade 75-octane gasoline, the fuel system fed by pressure from two compressed air tanks.

First V-1 launched on 13 June 1944 on London, and within a month, more than 100 per day were launched. In the June 1944 first successful salvo of V-1s launched against London, 73 out of 244 made it. Production was such that 614 V-1s had been manufactured in two weeks. By July 1944, 2,745 V-1 had been dropped on England, taking the lives of 2,752 people. Over 8,600 V-1s were fired at London, but only 2,400 arrived, the slow-flying CM being easy for fighter planes to shoot down. Nonetheless 5,500 people were killed, 40,000 wounded and 130,000 were homes destroyed and 750,000 damaged. 9,000 were also fired against various European cities, including 5,000 against Antwerp in a definitely military role.

The Allies had received reports of the new weapon, and photo-reconnaissance of the catapult-launcher, before they located the manufacturing site in November 1943, close to Peenemunde, which became a high-priority target. A massive strike by 1,300 bombers of the U.S. 8th Air Force followed.

The suicide kamikaze pilot of 1945 could be considered the second "cruise missile": a propeller-driven Zero fighter plane packed with HE that dive-bombed through a hail of flak onto U.S. Navy ships in the Pacific, aiming for the base of the engine exhaust funnels.

As usual, the lesson was not lost on the U.S., who adopted the cruise missile concept, added nuclear weapons, and a development effort that took three decades to perfect. It is the latest nuclear weapon carrier, having lain essentially dormant for many years until technology improved enough to make it an effective modern weapon in 1979. Because technology was not ready for the cruise missile, they struggled throughout the 1950s to build a usable CM, finally coming up with the Hound Dog CM in 1959, used as a stand-off Mk 28-carrying missile, with two loaded on the wings of a B-52.

180 Pershing II's and 464 GLCM (Ground-Launched Cruise Missiles) were stationed in Europe in the early 1980s before being withdrawn by the signing of the 1987 INF Treaty negotiated by Ronald Reagan. There were 108   Pershing II IRBM's and 96 GLCM's in West Germany, 160 GLCM's in England

(Greenham Common), 112 GLCM's in Italy, and 48 each in the Netherlands and Belgium, most or all of which had a strategic capability against Eastern Russia, up to Moscow.

The ALCM (Air-Launched Cruise Missile) was also the latest advancement to give new life to the Air Force's obsolete B-52 bomber, which became one of the cruise missile's carriers and launch platforms just outside the Soviet border – a "stand-off weapon", that replaced the SRAM missile, which was ostensibly for anti-radar site use.

Like the B-52, the ALCM, with a speed of 550 mph, is slow to reach its target, but its TERCOM radar makes it extremely accurate, and its very low flying altitude makes it hard to defend against. And the Tomahawk CM only costs $1.8 million dollars (in 1989; exclusive of warhead cost).

The modern Tomahawk SLCM, initially deployed in 1979, also gave new life to the U.S. Navy, which was no longer restricted in surface ships to carriers with a strategic nuclear capability. Instead of twelve or fifteen aircraft carriers for the Russians to track, there were now hundreds of ships capable of carrying the Tomahawk. [Gardiner, p. 139 (1993)]

## Early Cruise Missiles

J33 = Allison J33 turbojet

| | Span (ft.) | Length (ft.) | Weight (lbs.) | Payload (lbs.) | Power plant | Thrust (lbs.) | First Successful Flight |
|---|---|---|---|---|---|---|---|
| Matador (Martin) | 28.6 | 39.7 | 11,550 | 3,000 | J33 | 4,600 | Mar. 1949 |
| Regulus I (Chance-Vought) | 21 | 33 | 14,522 | --- | J33 | 4,600 | Mar. 1951 |
| Mace (Martin) | 28.6 | 46.8 | 12,750 | 2,920 | J33-A-41 | 5,200 | 1956 |
| Navaho (American) | 40.2 | 87.3 | 12,050 | 3,000 | 2 x R147 | 20,070 | Mar. 1957 |
| Snark (Northrop) | 42.2 | 67.2 | 49,000 | 7,000 | J57 | 9,500 | Aug. 1957 |
| Regulus II (Chance-Vought) | 20 | 67.2 | 22,564 | 2,920 | J79 (Afterburner: 15,000 thrust) | 9,600 | Nov.1957 |
| Rascal | 16.7 | 32 | 18,500 | | | | Nov. 1957 |
| Hound Dog (North American) AGM-28B | 12 | 42.5 | 10,147 | 1,742 | J52 | 7,500 | April 1959 |

[AD-A162 646 (1985)]

Matador MGM-1 (TM-61C) (1954 – 1962)

W-5 warhead

GLCM (winged surface-to-surface cruise missile)
Range:  620 miles
Speed:  >650 mph; Mach 0.9
Propulsion:  One Allison J33-A-37 Turbojet (producing 4,600 lbs. thrust), with Aerojet solid-fuel rocket
        booster (producing 55,000 lbs. of thrust for 2 seconds).
Dimensions:  54" dia. x 39' 8" long (with booster); Maximum wingspan:  27' 11"; Wing area:  176.7 sq. ft.
Weight: 13,000 lbs.
CEP:  2,700 ft.

Contractor:  Glenn L. Martin Co., funded by the U.S. Air Force

>1,200 produced, with the last of the operational missiles replaced in 1962. The Air Force's first operational guided, tactical missile.  Deployed in West Germany in 1954.  Also deployed in Taiwan and South Korea.  Evolved into the Mace missile.

"[U]nimpressive." [Neufeld, p. 128 (1990)]

Inspired by the success of the Nazi V-1, Goodyear came up with the precursor to (the later cruise missile guidance technology) TERCOM Radar navigation in 1948, the ATRAN (Automatic Terrain Recognition And Navigation) guidance system.  It was fitted to the Air Force's Matador and Mace Cruise Missiles.

Cancelled Nov. 1956, in favor of the Mace Cruise Missile.



Regulus I (RGM-6B) (1954 - 1964)

W-5 warhead

Short-range, submarine, surface cruiser, or carrier-launched; land-attack cruise missile
Dimensions:  54" dia. x 33' long; Wingspan:  21'
Range:  575 nm
Propulsion:  Air-breathing Allison J33 turbojet, with Aerojet solid-fuel rocket boosters
Speed:  Mach 0.9
Weight:  14,500 lbs.
Guidance:  Radio-controlled, command guided

Contractor:  Chance Vought, funded by the U.S. Navy.

514 built.  Retired with the introduction of the Polaris submarine in the early 1960's.  [Spinardi, p. 20 (1994)] [Gardiner, p. 217 (1993)] [Goldberg, Part 1, p. 431 (1981)]



Mace B (MGM-13B) (1955 - 1966)

W-28 warhead

Surface-to-surface missile
Dimensions:  54" x 46.8'; Wingspan:  22' 11"; Wing area:  151.5 sq. ft.
Weight:  18,750 lbs.
Range:  625 nm (Mace A); 1,150 nm (Mace B)
Speed:  Mach .8

Propulsion: Allison J33-A-41 turbojet (producing 5,200 lbs. of thrust) with one Thiokol solid-fuel rocket booster (producing 50,000 lbs. of thrust)
Guidance: Terrain Matching (Mace A); Inertial (Mace B)

Contractor: Glenn L. Martin Company

Over 1,000 produced. An improved version of the Matador missile. Mace A used a new guidance system known as ATRAN (Automatic Terrain Recognition and Navigation). Mace B was all inertial guidance. Deployed to Europe in 1959. Between 1966 and 1971, the Mace was replaced by the Pershing I missile.


Triton

Successor to Rigel CM. Funder: Navy; cancelled Dec. 1955 [Hansen, VII-272 (1994)]
Dimensions: 34.5" dia. x 34.5' long
Span: 140"
Range: 2,000 nm
Speed: Mach 3.5
CEP: 600 yards


Rascal (GAM-63) (1957) (cancelled December 1958; never operational) ["Rascal (B-63) Weapon System (1963)] [AD046 714 (1954)] [AD113 976 (1956)]

Carried a Mk 27 warhead. The warhead compartment was essentially a cylinder of 44" diameter, and 75" length. Airburst delivery.

Formerly code-named B-63 from Project MX-776B. Long range, stand-off, liquid-fuelled, rocket-propelled, strategic, supersonic, air-to-surface parasite missile. Launched from about 40,000' from a medium or heavy strategic bombers like the B-36, B-47, and B-52, in order of priority from highest to lowest. Predecessor to the Hound Dog missile on the B-52. Planned and tested inertial guidance system for Rascal cancelled. Used RCA magnetron with a Raytheon backward-wave radar tube in its laminated ogive radome nose section.

Dimensions: 4' dia. x 32' long; 200.5" rear canard cruciform wing span; also had 137.4" wide forward canard cruciform wings.
Weight: empty weight of 5,903 lbs.; propellant weight of 9,747 lbs.; pressurized (to 6,000 psi) nitrogen gas (for propellant flow) weight of 91.5 lbs.; warhead weight of 3,000 lbs.
Propellant: Kerosene (JP-4 aviation fuel; 295 gallons) fuel, and Inhibited Red Fuming Nitric Acid (IFRNA; 615 gallons) as the oxidizer, both in ring-stiffened cylindrical 6061 aluminum alloy tanks integral with the main cylindrical semi-monocoque fuselage (airframe missile body).
Maximum Speed: Mach 2.7
Thrust: 12,000 lbs. of thrust at an altitude of 40,000' to get to supersonic speed for its climb to cruising altitude, then a sustaining thrust of 4,000 lbs. to maintain the supersonic speed; the three identical 4,000 lb. drilled cast aluminum thrust chambers internally coated with 4 mil thick tungsten carbide (WC), each consisting of four 72-pair, stainless steel, impinging showerhead injectors
Range: 90 nm
CEP: 1,500' if launched from 75 nm

Contractor: Bell Aircraft Corp.

Regulus II (RGM-15) (1958)

Range: 1,200 nm
Speed: > Mach 2; in other words, supersonic, unlike the Regulus I.
Propulsion: GE J79 Turbojet with Rocketdyne solid-fuel rocket booster
Dimensions: 56.4" dia. x 56.9' long
Weight: 26,000 lbs.
Guidance: Inertial

Contractor: Chance Vought for the U.S. Navy

Canard-configured Cruise Missile. Submarine or cruiser-launched. Had to be launched from an unsubmerged submarine. Canceled in 1958 in favor of the introduction of the Polaris A-1 SLBM, so it was never deployed.


GLCM

Weight: 14,000 lbs.
Range: 1,200 nm
Cruising Altitude: 100 m
Speed: Mach .9
Propulsion: Turbofan
Guidance & Control: Inertial and Map-matching

Contractor: Thiokol

Began as an advanced version of the MACE, and a ground-launched version of the Tomahawk CM. The Automatic Terrain Recognition and Navigation (ATRAN, a trademark of Goodyear Aerospace Corporation) system was used on the Mace CM. The ATRAN system provided continuous missile navigation without the use of an INS, which at the time was not suitable in such a system because of cost and performance considerations.


Navaho (cancelled 1956) (follow-on 1960)

Supersonic, inter-continental (like Snark) air-launched with rocket and RAM jet engines, and a guidance system. "[U]nimpressive". [Neufeld, p. 128 (1990)]

Dimensions: 6.5' dia. x 87.3' long
Length with booster: 92' 1"
Wingspan: 42' 8" (13 m)
Weight: 120,500 lbs.; Weight with booster: 300,500 lbs. (136,000 kg)
Speed: Mach 3
Range: 5,500 nm
Altitude: 71,000'
Materials: Titanium alloy to resist Mach 2-3 speed heating
Guidance: Inertial (SINS)

Contractor: North American Aviation

Development led by William Bollay who added two ramjets to the wing-tips of the liquid-fuelled rocket engine propelled missile, building on the last designs of the German V-2, which had wings added to it to increase its range, during the final months of the war.  The rocket engine started at lift-off, and took the CM to nearly Mach 3.  Flight testing began in 1956, but the Navaho was canceled the same year, in favor of the Snark, and IRBM and ICBM development.  [Heppenheimer, p. 184-185, 198 (2004)]  Originally going to carry a 10 MT bomb.  [GACAEC, 50th, p. 43 (1956)]


Hound Dog (AGM-28B) (May 1960 - 1976)

1.44 MT W-28 warhead

Air-to-surface missile.  Canard-configured.  Delta wing design.
Dimensions:  28.5" dia. x 42' 6" long; Wingspan:  12' 2"
Weight:  9,600 lbs. (4,350 kg)
Speed:  Mach 1.6
Range: 200 nm (low altitude cruising) - 500 nm (high altitude launch) (965 km)
Propulsion:  Air-launched, air-breathing turbojet by Pratt & Whitney (J52-P-3 motors) of 7,500 lbs. thrust.
CEP: >1 nm
Guidance:  Inertial-guidance by North American Autonetics, supplemented by a star-tracking system by
          Kollsman Instrument Co.

Reached a peak inventory of 593 in service with SAC in 1963; over 300 deployed as late as 1975.  Considered of "low reliability."  Replaced by the SRAM missile, when it was developed.

Two air-launched Hound Dogs carried by the B-52G/H (on its wings, on underwing pylons) as a stand-off weapon (fired well outside of the Soviet border) (the Hound Dog was could be launched from the B-52 when it was 500 miles from its border target) for destruction of upcoming Soviet air defenses (early warning radar sites and fighter plane airfields near the border) and allow the bomber's penetration.  Was inherently hard to EMP pulse due to the continuous metal skin around the missile and the existence of multiple connecting cables between electronic boxes that diluted the EMP signal.  ["History of Strategic Air Command January – June 1968" (1969)]


Snark (SM-62) (March 1960 – June 1961)

Dimensions:  60" dia. x 806.44"
Wingspan:  42' 3" (20.5 m)
Weight:  55,000 lbs.
Range:  6,300 nm (Intercontinental)
Speed:  Mach 0.9 (i.e., subsonic)
Propulsion:  Pratt & Whitney turbojet, with two Aerojet solid-fuel rocket boosters
Altitude:  48,000'
CEP:  8,000 feet; 1 mile accuracy at 5,000 mile range
G&C:  Inertial/Star-tracking Stellar navigation

A cruise missile, it was the first intercontinental missile.  "[U]nimpressive".  [Neufeld, p. 128 (1990)]  Smaller, but more advanced than its cancelled brother, the also intercontinental, Navaho.  Slower than the Navaho, but it had a low-level target approach, and was a small radar target. [AD-A328 596 (1997)] Year (briefly) in service then cancelled in mid-1961, with introduction of SLBMs, and the introduction of the Atlas and upcoming Titan I ICBMs. [Goldberg, part 2, p. 872 (1981)] ["SAC Missile Chronology 1939 – 1988" (1990)]

Gryphon (BGM-109G) (Boosted Guided Missile) (1962)

A GLCM

Dimensions:  1' 9" dia. x 18' 2" long; with booster 20' 6" long; Wingspan:  8' 7"
Weight:  2,650 lbs. missile alone; 3,200 lbs. for missile with booster
Speed:  approximately 500 mph
Propulsion:  Williams International F-107-WR-102 turbofan engine, producing 600 pounds of thrust; solid
          fuel rocket booster produced 7,000 lbs. of thrust
Guidance:  The guidance system is provided by McDonnell Douglas, and consists of inertial
          guidance and updates using the TERCOM system.
Range:  1,500 miles
CEP:  100 ft.

464 GLCMs planned for deployment around Europe; only 288 actually deployed in 5 European countries.
[ADC 021 800 (1980)]  [AD-A258 351 (1992)]

Yet these early cruise missiles were all canceled, having reached maturity just at the approximate time of ICBMs, and in their early form, clearly inferior.  The cruise missile would have to wait for a more propitious time to reenter the arsenal, as an effective tool of accurate nuclear destruction.

The Navy's TALOS-W SAM missile, powered by Ramjets, was really the only cruise missile-like weapon system that achieved success, with full deployment.  The BOMARC also achieved success.  The Hound Dog was deployed.

Not all the development work went to waste however, as designers and engineers used appropriate bits and pieces of the better designs in the continued missile work, including INS subsystems, and Titanium alloys, and especially TERCOM guidance in the modern Cruise Missile that reared its head fifteen years later.


**Modern Cruise Missiles and Their Warheads**

The turbo-fan or fanjet jet engine first made its debut with the Boeing 747 in 1968, which was equipped with four of the more fuel efficient engines.  Though it was originally invented in 1943 by the Germans, the first production turbofan was built at Rolls-Royce.

There is the turboprop (propeller) aircraft, and the more modern turbojet (jet) aircraft.  The turbofan is a combination of the two.  A propeller is mounted on the front of the airplane, attached to the jet's turbine.  Air flows partially into the jet, and is partially ducted around the central turbojet.

50 - 80% of the air flows around the jet depending on whether it is a low-bypass or high-bypass turbofan engine, respectively.  The turbofan is much more fuel-efficient (15 - 20% less fuel consumption) and quieter than the turbojet, with a lower IR signature and is optimized for low-altitude flight.  In 1968, the use of high-performance fuels like Shelldyne H (mixed with the thinner, methyl cyclohexane, to lower its viscosity) promised a 50% range improvement over the standard JP-4 and JP-5 aviation fuel.

It is most efficient at the 300 - 600 mph range.

In other words, it's perfect for propulsion of the cruise missile, with further range and counter-measure-proof, especially with low level flight.

The turbine blades of a turbofan engine are subject to high heat and stress, and are manufactured of single crystal metal. Alloys in widespread use are nickel super-alloys. The largest manufacturers of turbofan engines for both military and civilian markets are, in order, GE, Rolls-Royce, and Pratt & Whitney.

The development by the Air Force Aero Propulsion Lab in the late 1970's of JP-9 kerosene for Cruise Missiles, allowed for an increase of 20% more range for the same volume of fuel, though they adopted the much lower cost JP-10 as the standard fuel for cruise missiles. [AD-A528 970 (2005)]

For guidance, first there was the development of Synthetic Aperture Radar (SAR), by the Sensors Directorate of the Air Force Research Laboratory at Wright Field, Dayton, Ohio. The development of SAR has been described as "probably the most remarkable development in military sensor technology in the 20th Century". [AD-A528 970, p. 310 (2005)] SAR is built upon another revolutionary development, phased array radar, and is an airborne antenna used to simulate a much larger antenna by collecting radar data as the cruise missile moves forward, and then processes the information as if it came from an antenna several hundred meters long. This results in vastly increased resolution and the ability to penetrate clouds, foliage, and camouflage.

(Phased Array Radar is radar with no moving parts, such as the previous rotating dish antenna. Instead, PAR has several hundred to several thousand transmit/receive modules than can be electronically actuated to scan the target, as if it were moving physically.)

The TERCOM (Terrain Contour Matching) radar guidance system was first patented by LTV (Chance-Vought) in 1958. Development of the TERCOM system, which matched a pre-stored profile of terrain altitude with returns from the missile's radar altimeter, was the biggest stumbling block and presented the greatest difficulty in building the modern cruise missile. It required the development of the microprocessor, around 1970, and semiconductor RAM memory that enabled the hardware for TERCOM. Along with the technique of the use of the Kalman linear filtering algorithm (developed by Dr. Rudolph Kalman), which resulted in reduced memory use (16K), the stage was set.

In the system, a terrain map is divided into a matrix of squares ("cells"), each cell 100 ft. to 3,200 ft. to a side. The LTV system consisted of 64 cells, each 400 feet to a side yielding a 4.9 nm strip map.

Engineers assign each cell an average elevation from contour maps or satellite mapping reconnaissance systems. During flight, a radar altimeter measures the average elevation, and then at checkpoints compares the values with a digital map stored in the computer's memory.

The problem with TERCOM was that the necessary database of digitized terrain profiles of Soviet territory did not exist. Between 1968 and 1974, using radar reconnaissance satellites this was rectified, since Soviet overflights by the U-2 had been banned since the Gary Powers shoot-down.

TERCOM requires a radar altimeter to measure ground clearance and a barometric altimeter to measure altitude above sea level, and was chosen in 1975 by the DoD for the Navy Tomahawk, the ALCM, and later the ACM.

TERCOM is mated with an Inertial Guidance System to navigate to the first checkpoint, and to reach land from the open sea of sea-launched cruise missiles (SLCM & ALCM & ACM). Since accuracy equal 0.4 times cell size, the cell size is reduced as the CM approaches the target. The open literature claims an accuracy of 100 to 600 ft. with 165 ft. accuracy reportedly demonstrated in 1960, significantly better than ICBMs, with the limitation of the CM being a "slow flyer" taking hours to reach its target, unlike the 30 minutes of ICBMs.

A quasi-TERCOM system was first disclosed in U.S.P. #3,071,765 (1963; filed in 1950). As well as U.S.P. #4,163,231 (1979; filed in 1968), and finally U.S.P. #4,993,662 (1991, filed 1970).

The power line avoidance radar system was disclosed in a patent to Texas Instruments. [U.S.P. #3,568,187 (1961; declassified 1971)]

"A TERCOM *map* is simply a rectangular array of digital terrain elevations, above Mean Sea Level (MSL), located along the predicted flight path of the cruise missile. It is the portion of terrain over which a correlation will be done with sensed data to determine INS error. Each elevation value represents an average for a particular square cell. These cells may vary in size. For example, a 300 foot cell would contain the average terrain elevation value above **MSL** within a square area whose edge is 300 feet long (and whose area is 90,000 square feet). The cell sizes used may vary during the course of the mission depending upon the desired accuracy and reliability, and on a **1%** missile computer storage capacity.

A TERCOM update area is a TERCOM map, or map set, prepared by DMA that has successfully passed through an evaluation procedure to ensure its quality and is suitable for use in the cruise missile guidance system. The test strip or column of data is the one-dimensional (downtrack) profile generated by the forward motion of the cruise missile, coupled with the operation of the downward oriented radar altimeter. The term "terrain matching" refers to the "correlation-like" process whereby the reference map (stored in the missile's on-board computer) is compared with the "live" sensor data to determine the location of best match. This information is then used to update the state of the Kalman filter within the guidance system to help correct errors resulting from imperfect gyroscopes and accelerometers. Hence, TERCOM is a terrain (or map) matching guidance updating system. Each TERC&M reference map comprises a given number of cells in the down-track and across-track directions, determined in part by the expected INS errors (hence missile position error) at a given point within the flight and the characteristics of the terrain matching correlation algorithm used. Each TERCOM reference map also has a given azimuth orientation relative to a true north heading to facilitate the mission planning process.

In the cruise missile, a guidance updating system is needed to aid in the removal of time-varying error sources within the INS (primarily A because of gyroscope drift and accelerometer bias). In the TERCOM system, a terrain profile computed inflight from barometric/inertial and radar altimeter data is compared with one determined beforehand and stored in the on-board computer.5 (This predetermined data is the DMA TERCOM map). A correlation-like algorithm (Mean Absolute Difference) is then used to determine in near real time the position of "best match" between the down-track (one-dimensional) altimeter test data and the two-dimensional computer stored reference map. The difference between the location of "best match" and the vehicle's estimated location (derived from the INS) provides an estimate of the down and across-track position errors that are present at the time of the update. The information is then utilized in the missile's on-board Kalman filter not only to introduce a flight path correction to "steer out" the errors, but to reduce their time-varying growth during the course of the flight. The net result is that the INS error accumulation may be substantially reduced during the course of a flight through several set, of TERCOM updates.

During the course of the cruise missile flight, different sizes of TERCOM maps are utilized to account for the variations in INS accuracy, algorithm performance, and on-board computer storage capabilities. Thus, for the initial or landfall update, fairly coarse cell sizes can be utilized because minimum flight error is not as critical as ensuring a reliable update to help remove potentially large position errors. These could be up to several miles in cases where initial position and velocity values from *the* launcher or carrier were poor or there is a long standoff launch range over water (where TERCOM is not applicable).  The en route, mid-course, and terminal maps utilize data with progressively smaller cell sizes, because the missile guidance system error should have been reduced by previous updates and to provide increasingly more accurate estimates of the INS error characteristics.  To permit estimation of the down and across-track INS velocity errors and to increase overall reliability, most TERCOM updates utilize three individual correlations performed in succession. Because a non-Doppler system is used, velocity errors cannot be estimated from a single correlation. Given the importance of this component in the resulting cruise missile position error as it propagates with time (even if zero initial position error existed), a velocity error estimation technique--two or more successive correlations--is needed to compensate.  When these multiple correlations are coupled with a voting logic (e.g., two out of three correlations must match to have a valid TERCOM *update),* the probability of obtaining a false update can often be substantially reduced.

The **DMA** prepares each candidate TERCOM reference map from a digital matrix of terrain elevation data. The matrix location is determined by terrain roughness and uniqueness considerations coupled with mission planning considerations. The objective is for the map to meet mission planning constraints and to have a high probability of correct correlation **(PCC)** value.

Given a terrain elevation matrix of suitable roughness and uniqueness, a number of computer programs and subroutines are used to ensure that the terrain selected for a reference map will support proper TERCOM operation. First, an interactive computer program called STAT prepares candidate reference map files, calculates terrain roughness statistics, and presents abbreviated AUTOMAD results. STAT assists in quickly identifying and evaluating the most likely sites for the desired maps. The AUTOMAD computer program is the central routine for the map selection and validation process. It performs all TERCOM correlation operations done in flight. There is no true measured altimeter profile available; therefore, AUTOMAD uses the reference terrain profile itself, a more complex and costly Monte Carlo simulation is avoided, and a considerable amount of time and money are saved. The results achieved during flight testing have demonstrated a high degree of confidence in the PCC values computed by the AUTOMAD program.

The complexity of this simulation makes it necessary to validate the individual sub-models present by extensive flight testing. This is typically performed by means of tests in aircraft equipped with a Cruise Missile Guidance Set (CMGS) to reduce overall program costs. Flight testing is also performed to evaluate the suitability of candidate changes to the TERCOM system (i.e., altimeters, and terrain elevation data obtained from different sources), as well as to validate the performance of the system over operationally representative terrain.

## Terrain Following

To enhance survivability, land-attack cruise missiles can use a low altitude terrain following mode to minimize the probability of detection when over hostile territory. A safe terrain following clearance Above Ground Level (AGL) is determined for each leg of the flight during mission planning by running missile simulations over Digital Terrain Elevation Data (DTED) and Vertical Obstruction Data (VOD) along the route. These "clearance plane settings" above the terrain are then preset into the missile computer along with other mission commands prior to flight. In flight, the missile enters and departs the terrain following legs using altitude information from the radar and barometric/inertial altimeter subsystems. Once in the terrain-following mode, the missile attempts to maintain the desired clearance plane altitude using only the radar altimeter data' and appropriate throttle and pitch controls.

The data used in the terrain following process and in the clobber analysis7 module simulations during mission planning are an overlay of the DTED and VOD files. The DTED is primarily oriented toward natural terrain elevation characteristics, and that in the VOD file includes man-made features whose AGL height is greater than some prescribed value." [ADA141082 (1982)]

ALCM, AGM-86B – Air-Launched Cruise Missile (December 1982)

Warhead: W-80-1

Dimensions: 24.5" dia. x 234" long
Wing Span: 143.6"
Wing Area: 11 sq. ft.
Wing Sweep: 25°
Fuselage Cross-Section: triangular
Weight: 3,175 lbs.
Range: 1,500 miles
Speed: 500 mph (800 kmph) (a low-flyer: flies at a maximum of 500 feet/150 m off the ground to avoid radar)

Propulsion: F107 turbofan engine, weighing 66 kg and with a thrust of 600 lbs., by William International
Company, Walled Lake, Michigan
Fuel: JP-10 ($C_{10}H_{16}$), to provide the needed low freezing point [AD-A111 686 (1981)]
Navigation/guidance: P-1000 platform P-1000 platform; LC 4516-C computer LC 4516-C computer; LCM
9000 memory (32k) [AD-A141 082 (1982)]

Contractor: Boeing

Air-to-surface strategic cruise missile. The first, and easiest of the modern CM's to design, as it was launched from a flying B-52, and so didn't need ground take-off booster rockets. Production of a total of 1,140 nuclear-armed CM's was completed (December 1982 – 1985) before production was stopped in favor of the advanced ACM in 1985.

Equipped with a 3.5 kW generator, cooled by ram air, and used a 28 vDC system. The thermal battery is of the Ca/CaCrO4 type. [AD-A174 324 (1986)]

Like most modern CM's, the exhaust temperature is a low 600 °F, helping defeat IR detection of the CM. [AD-A257 717 (1992)]

The B-52G & B-52H can carry twelve ALCMs in external racks, and some B-52H's can carry eight more internally as well in the SRAM launcher, for a total of twenty W80's. Trapezoidal fuselage shape and elevron controls.

INE will be used to designate the inertial guidance platform, digital computer, and power supply for the land-attack GD/C and Boeing cruise missiles. An INE, together with a radar altimeter and chassis, constitutes a Reference Memory Unit and Computer (RMUC). When integrated into a package, the RMUC is known as the Cruise Missile Guidance Set (CMGS), which is provided by MDAC for the GD/C Tomahawk land-attack cruise missile. Boeing takes the INE, provided by HDAC, and adds a separate autopilot and associated computer and radar altimeter. The result is a distributed navigation/guidance subsystem that is incorporated in their ALCM (AGM-86).

Tomahawk SLCM (Ship or Submarine Sea-Launched Cruise Missile; also Air or Surface-Launched) BGM-109A (TLAM/N or TLAM-A): Tomahawk Nuclear Land Attack Missile) (June 1984)

Warhead: W80-0; 270 lb. warhead; 100 lb. guidance set [Mickelson (1979)]

Dimensions: 20.4" dia. x 18.25' long (w/o solid-propellant take-off booster); 20.5' long (w/booster)
Weight: 2,584 lbs. (w/o booster); 3,200 lbs. (w/ booster)
Wing Span: 8.75'
Wing Area: 12 sq. ft.
Wing Sweep: 0°
Fuselage Cross-Section: Circular
Speed: 550 mph; high subsonic (Mach 0.7 - 0.85)
Cruising Altitude: 50-250 m; 20-1,000 m
Range: 1,350 nm flight distance
Propulsion: J402 turbofan engine by TCAE, Toledo, Ohio, and Williams Research; Weight 60 kg and has
600 lbs. thrust
Fuel: RJ-4 ($C_{12}H_{20}$)
Accuracy: 12 - 50 m
Guidance: McDonnell Douglas TERCOM and DSMAC
Navigation/guidance: P-1000 platform P-1000 platform
LC 4516-C computer LC 4516-C computer
LCM 9000 memory (32k) LCM 9000 memory (64k)

Because "the best currently-used inertial guidance systems tend to 'wander' up to 900 meters off course for every hour of flight time," TERCOM is used. [AD-A257 717 (1992)] By August 1987, 464 TLAM/N's had been deployed all over western Europe, including the U.K. and West Germany. It was also installed in the 6 missile rotary missile launcher in the B-52.

Contractor: Raytheon Missile Systems, Tucson, Arizona [Raytheon Missile Systems (2002)] [AD-A141 082 (1982)]

GLCM (Ground-Launched Cruise Missile) BGM-109 (December 1983)

Warhead: W84 of supposedly 80 kt yield

Dimensions: 1' 9" dia. x 18' 2" long; with booster 20' 6" long; Wingspan: 8' 7"
Weight: 2,650 lbs. missile alone; 3,200 lbs. for missile with booster
Speed: 550 mph/880 kmph
Propulsion: Williams International F-107-WR-102 turbofan engine, producing 600 pounds of thrust; initially launched by its solid fuel rocket booster which produced 7,000 lbs. of thrust
Guidance: The guidance system is provided by McDonnell Douglas, and consists of inertial guidance and updates using the TERCOM system.
Range: 1,500 miles/2,500 km
Height of attack: Flight altitude of 15 - 30 m above ground
CEP: 100 ft.

464 GLCMs planned for deployment around Europe; only 288 actually deployed in 5 European countries. [ADC 021 800 (1980)] [AD-A258 351 (1992)] The GLCM was the follow-on of the Martin TM-81 Matador, the Martin TM-76 Mace, and the 1962 General Dynamics BGM-109G Gryphon.

ACM (Advanced Cruise Missile) AGM-129A (1986)

The ACM, where it stands today, is the most highly advanced cruise missile ever developed. "Stealthy" air-to-surface cruise missile; replaced the ALCM.

Dimensions: 29" dia. x 250" long; Triangular in cross-section with rounded edges, with a height of 25", and a base width of 27.6"; sharp nose for low frontal RCS.
Weight: 2,750 lbs. (1,250 kg)
Wingspan: 122"; forward swept wings
Engine: Williams F112 turbofan; low bypass ratio turbofan; thrust: 732 lbs.; length: 31"; weight: 161 lbs.
CPU: With 128k RAM, and 64k ROM
Guidance: Inertial: four gimbal-tuned rotor gyroscopes. Also uses lasers to map the ground image, creating its excellent CEP and increasing its stealthliness.
CEP: 13 m

Contractor: General Dynamics and Hughes Aircraft, then later, Raytheon
Number: 468 (1987 - 1993) at $3.8 million each (1991 dollars)

The B-52H can carry 6 beneath its wings, plus 8 on the internal (revolver-like) rotary launcher.

The ACM is specially shaped so that its leading and trailing edges reflect radar pulses away from their source. It is a third generation stealth weapon, and reduces the radar cross section from less than one-tenth to less than one-hundredth of a square meter in the forward sector ["Soviet Work on Radar..." (1984)] However, the ACM appears clearly to radar from above it in altitude. [Clearwater (2006)]

It is covered with a low reflectance paint (for lowered IR and visual signature), and radar absorbent material (RAM) made of a hard rubber-like material impregnated with ferrite spheres dispersed throughout. The ferrite scatters and diffuses the radar pulse's energy into the rubber dielectric material which absorbs it as a "lossy dielectric". [U.S.P. #5,717,397 (1998)] [U.S.P. #6,662,546 (2003)] ["Air Force Stealth Technology Review" (1991)] Kevlar 49 & silicon carbide are also powerful radar/microwave absorbers.

Alternatively, magnetizable, spherical iron particles (0.5 - 20 microns (μm) in diameter) in a dielectric (non-conducting) matrix of silicone thermosetting polymer. Superior to previously used ferrite particles, which don't work at frequencies above 30 GHz. 40 mil layer attenuates the enemy radar signal 12 - 20 dB. 6 kg iron powder with 700 g monomer. Catalyzed to form polymethylsiloxane and polyphenylsiloxane. [U.S.P. #4,173,018, (1979)]

*   *   *   *

The first nuclear Triad of 1962, the Polaris SLBM, the Minuteman I, and the B-52 was now upgraded with the sea-launched cruise missile, SLCM, the Trident II, and the Peacekeeper, the new First Strike Triad was completed, the ultimate in very expensive nuclear striking power.

To make sure the Air Force – a redundant, useless capability -- was able to stay feeding at the pig trough – their obsolete B-52's were equipped with air-launched cruise missile versions. Thus, the B-52 no longer had to risk crossing the Soviet border, flying "down on the deck", a maximum of 200 feet above ground to evade Soviet radar and missiles that had made life too difficult for the B-52 in the 1980s. [Hadley, p. 230 (1987)]

A submarine directly off the coast of USSR launches its ground-hugging cruise missiles, which fly under the radar, at 100-300 feet, initially steered by gyroscopic INS until they are over land, and then, equipped with TERCOM radar hit their targets with pin-point accuracy, using a third system, the DSMAC (Digital Scene Matching Area Correlator), which compares the target scene with its image stored in memory.

The slowness of CM's is compensated by their ability to fly at tree-top level and equipped with stealth technology, for evading radar detection, and for a surprise attack.

With IRBM Pershing II's and GLCM's gone from German and U.K. deployment with the INF Treaty signed Dec. 1987), SLCM's were an able substitute, not subject to treaty rules. With SLCM's available, the Pershing II's were superfluous.

And to complete the package with ultimate versatility, the Tomahawk cruise missile could also be equipped with an HE warhead, instead of its nuclear warhead.


**Intelligence Collection: From Airplanes to Spy Satellites**


"Actually, your Hubble [tele]scope is a Keyhole-class satellite."

-- AF intelligence officer to Chaisson
"The Hubble Wars" (1994)
Eric J. Chaisson


First there was PHOTINT (photographic satellite intelligence, also called IMINT, image intelligence), then ELINT (electronic satellite intelligence; radar and missile telemetry), COMINT (communications satellite intelligence; radio communications), and SIGINT (signals satellite intelligence; ELINT and COMINT together). They have largely supplanted the dangerous and difficult work of HUMINT (human intelligence; human spies) for the more expensive, but accurate and versatile, hardware solutions.

**The U-2**

The U-2 was built at President Eisenhower's request to find out the true nature of the closed ("behind the Iron Curtain") USSR's nuclear capabilities, about which they knew almost nothing. The U-2 sub-sonic jet, black-painted, reconnaissance plane was built for long flights at the high altitude of 70,000 feet, to evade the Soviets' extensive air defense systems.

It was primarily used for searching for and photographing intercontinental bomber air-fields, uranium enrichment plants, plutonium production reactors, and (after 1957), most importantly, ICBM launch-pads (like Semipalatinsk).

Its camera's lens was 36" in diameter, and weighed 450 lbs. It was designed by Dr. James Baker, a renowned astronomer and lens designer at Harvard University. Instead of a lens to look skyward, he designed one to look from above, earthward. The lens was built by Perkin-Elmer Corp. of Norwalk, Connecticut. Computers were used for the first time for grinding the lens, instead of by hand. An MIT computer was used. The new DuPont Mylar-based thin, tough, plastic film was used as a base layer for the camera's film, developed by Kodak, and called Estar. Its use allowed large quantities of film to be rolled onto a spool for the U-2 project, and make longer, more productive missions possible.

The U-2 high altitude reconnaissance airplane was designed under aeronautical genius Kelly Johnson of Lockheed's secret California "Skunk Works" department, and had a wide wingspan of almost 71 ft., and a light weight of 11,700 lbs., fueled (and just over 10,000 lbs. unfueled), and could fly for 5.5 hrs. at the slow cruising speed of 460 mph, in the early model. Its wings weighed only three pounds per square foot, one-third the normal weight, and used a special low vapor pressure (low volatility) kerosene fuel, LF-1A, due to the high altitude. Regular JP-4 kerosene airplane fuel froze, or boiled off, or flamed-out at the high altitude, so a low boiling fuel was required.

U-2 flights typically took off from Turkey and landed in Norway: a 2,500 mile USSR overflight. Its 24 inch focal point camera could take 165 18" long shots, for a total of a 250' long spool of film. Pakistan and Japan were other take-off sites.

The U-2 photo-reconnaissance plane made its first flight over the USSR in 1956. Russian overflights came to an abrupt end in May 1960, when pilot Gary Powers, who had taken off from a military airfield near Peshawar, Pakistan, was shot down near Sverdlovsk (just north of Chelyabinsk-70, a secret nuclear weapon design facility – founded as a rival to Arzamas-16 – in the Ural Mountains, further north of the Aral Sea).

Powers' plane was brought down by a newly improved Russian SAM missile, the SA-2, improved just for that purpose, on orders directly from Soviet Premier Khrushchev. He had been enraged by the overflights' violation of Soviet sovereignty, and the embarrassing Soviets' impotence to do anything about it.

Powers had been provided by the CIA with a suicide-before-capture pin needle, a prick of which would kill him. However, he declined to use it. President Eisenhower assumed he had, and thus lied about the U-2 downing, being shot down himself, when Gary Powers, and his full confession, was then paraded gleefully by Kremlin officials in front of the international press. Eisenhower had been setup by the Russians, who had initially kept mum about their findings of the plane, its equipment, and its pilot.

A 1962 negotiated spy trade for the imprisoned Russian Colonel Rudolf Abel eventually freed Gary Powers from Soviet imprisonment, followed by repatriation.

And U-2 pilots for overflight of other countries, like China and Cuba, started to wear Air Force uniforms, unlike before. Powers could have been legally executed by the Soviets as a spy, not covered by the Geneva Convention, because he had been wearing civilian clothes.

**The Lockheed SR-71A Blackbird**

> "[T]he sky is a deep dark blue.  The sun is a big glowing
> globe in the blackness.  Cities are sparkling jewels in
> the black.  Everything is either sunlit or deep in darkness."
>
> -- "Flight" (2004)
> T.A. Heppenheimer
> an SR-71A pilot describing
> flight at 85,000' (26 km)

The follow-on of the U-2, the SR-71 flew from December 1964 to 1989.  It was retired from cruising high over the Soviet Union, when they finally proved they could shoot it down in a "demo" in 1986.

Cruising speed Mach 3.3 (over 2,200 mph/ 3,540 kmph) at 85,000' (25.8 km).  Maximum altitude of 100,000'.  Specially fueled by JP-7 jet fuel.  Twenty-nine operational SR-71A's were produced.  It was made of titanium, which though expensive, was twice as strong as aluminum, and could resist the heat of air friction, which heats the metal skin to 550 - 600 ℃.  Or possibly silicon carbide-coated titanium, which is doubly as heat-resistant as titanium.

It was powered by a pair of Pratt & Whitney after-burning J58 turbojets, twenty feet long, weighing 6,500 lbs., and producing 32,500 lbs. of thrust.  When it reached a certain high speed, the turbojet turned into a more powerful ramjet.

The ramjet used a movable conical "spike" in the jet's forward air intake, that regulated the speed of the air entering into the engine by moving forward to make it faster.  The spike had a movement range of twenty-six inches forward or backwards.

The SR-71 gathers PHOTINT, radar and radio signals.  Like the U-2, its predecessor, the SR-71 was designed under Kelly Johnson at his Lockheed "Skunk Works" facility and had a 93% titanium alloy airframe.

In 1986, was intercepted (but not fired upon) by six Russian MIG-31's within the range of their air-to-air missiles, and its career was over, as far as photo-reconnaissance went over the USSR.

First "stealthy" aircraft by its shape, and black IR-absorptive paint covering, which contained millions of microscopic iron balls in a dielectric matrix, that absorbed radar waves, it kept its radar cross section very low.  Supersonic speed also dramatically reduced the chance of detection by radar. [Nijboer (2003)] [Long (2001)] [Richelson, p.20 (2001)]

The SR-71 was considered in 1971 for post nuclear strike reconnaissance of the USSR ["A Study of SR-71..." (1971)], but such use was eclipsed by the DSP Satellite Constellation which arose in the same timeframe.

**Reconnaissance ("Spy") Satellites**

> "With the best possible diffraction-limited camera at an altitude of 120 miles,
> two objects can't be closer together than a few inches and still be recognized
> as distinct. This resolution is adequate to distinguish aircraft or tank types

and to count soldiers, but not to read license plates, assuming we haven't yet orbited any optically perfect ten-foot mirrors." [Alvarez, p. 226 (1987)]

In 1903, Russian scientist Konstantin Tsiolkovsky, detailed the mathematics of orbiting in space an artificial satellite.

The idea of reconnaissance (spy) satellite for intelligence gathering no doubt came to mind with the seminal 1946 report "Preliminary Design of an Experimental World-Circling Spaceship", a now classic RAND report. In 1953, the CIA commissioned a RAND report on the use of satellites for spying. The secret report produced, called Project FeedBack was positive, envisioning satellites for photo-reconnaissance, radio transmissions, and IR for anticipated ICBM missile launches.

Eventually, spy – surveillance – satellites would include versions for optical photography, infrared, magnetic, gravimetric, radar imaging, and SIGINT (RF communications, radar sites, etc.), and navigation and missile targeting (Navstar – GPS, the Global Positioning System).

However, the first satellite ever launched was Sputnik I, in 1957 by the Russians. It was only a simple satellite, an aluminum sphere weighing 184 lbs. with projecting antennae, and emitting a radio-frequency "beep" that could be picked up around the world. And it showed the U.S. that Russia was now capable of launching nuclear weapons from sub-orbital trajectory Inter-Continental Ballistic Missiles (ICBM's) at it.

The space race was on, and as well as a new phase in the arms race.

Using the name "Discoverer" as cover for its classified use, 38 satellites were orbited in the Discover series by the U.S., many of them failures. The Discoverer I was launched in January 1959. Discover XIII was the first successful film ejection and recovery in the ocean.

Discoverer XIV, launched in August 18, 1960 – Corona Satellite Mission 9009 – was the first successful film-return, as all the bugs were ironed out. An 84 lb. gold-plated drum containing twenty pounds of film was ejected from the satellite, and slowed by a parachute, was picked up in mid-air by an Air Force cargo plane. After Discoverer 38, launched in February 1962, future launches were classified and thus not announced publicly.

The real classified name for Discoverer was the Corona KH-1 (Keyhole), a photo-reconnaissance satellite, and the first spy satellite. It had a ground resolution enabling it to see and photograph 75' ground objects [National Reconnaisance Office (1973)], using a 5' long camera with an 18" diameter lens, and 24" focal length (the Corona series, the KH-1 to KH-4B, 144 satellite launches from 1959 to 1972, all had 24" focal lengths). The initial Corona satellite weighed only 92 lbs. with the film weight an additional 53 lbs.

This obviously (since it was a high altitude satellite versus a much lover level airplane) compared unfavorably with the U-2's 2.5 feet resolution, and stereoscopic, as opposed to the first satellites' monoscopic view, but such is life. Lockheed had overall responsibility and construction of the upper stage of the rocket.

It ejected an 84 pound capsule, recovered in mid-air by precision flying, containing a minimum of 6,200 photographic frames, each frame showing an area of 115 square miles in extraordinary detail. The satellite orbit had a perigee of 177 km (96 nm) and an apogee of 803 km (434 nm). It revealed that there was no "missile gap" with the Soviets, who had few ICBM's compared to the build-up of the U.S, who had been fooled by Soviet propaganda. They concentrated on areas off main railway lines and spurs, as more likely to have launching sites – a good, and accurate, bet.

Even worse, there was no "bomber gap", an earlier Soviet propaganda effort: they were putting all their limited financial efforts into their ICBM program, with only a token amount in their bomber program – a less effective nuclear threat. The U.S. bomber buildup of B-47 and B-52's had been a waste of money.

As well, their extensive nuclear anti-aircraft program had been a complete waste: the W-25 air-to-air, and W-31 Nike-Hercules SAM programs had been for naught.

By 1962, the KH resolution was reduced from 30 feet to 10 feet.

Within eight years and a massive budget, the resolution of the final version of the Corona (KH-1, KH-2, KH-3, KH-4A, and KH-4B, 95 successful), the KH-4B PHOTINT satellite, and operating between 1967 and 1972, was down to 1.5-3.3m (5-10') ground resolution. Approximately 10' in dia. x 49' long, and weighing approx. 25,000 to 29,000 lbs. It has an apogee of 150 miles, a perigee of 110 miles, and a 6 month life-time. Contractor: Lockheed.

From 1962-1964, the KH-5 ARGON geodetic satellite was also a wide swath, area surveillance system, provided accurate, high-resolution mapping data of the Soviet Union, with a 3" focal length mapping camera, for U.S. ICBM nuclear targeting. Along with the KH-6, which was instead a close look capability, they were both second generation models.

In 1964, the KH-6 LANYARD, orbited as low as 76 m (122 km), with a 66" camera. The best resolution it achieved was six feet, rather than the 2 feet planned. The LANYARD panoramic camera system was supposed to weigh 635 lbs. About 78 lbs. of film were carried for the main camera.

The Samos E-5 (plus the related LANYARD system) was a photo-reconnaisance satellite with a stereo camera with focal length 66", altitude of 180 nm, a ground resolution of only 5 feet, an f/5.0 aperture, and a film size of 5" width by 250-500' length. The 20° panoramic arrangement provided coverage of a ground swath 12 miles by 65 miles on each side, from 180 nm orbits, with the resulting strip of exposed film measuring 4.5" x 23". The E-6 had considerable advances in optics, vehicle stabilization, and camera mode technologies. [A-DA606 621 (1973)]

The KH-7 Gambit and KH-8 satellites, first launched in 1966, were the third generation of the Corona. The KH-8 was "spotting" as opposed to "area" photo-intelligence satellites, like the KH-7: they were close look satellites. They were low earth orbit satellites, the KH-8 as low as 82 miles. The KH-7 frame covered only 120 sq. miles, as opposed to 1,075 sq. nm for the KH-4. The KH-7's camera's focal length was 77 inches, and the film width was 9.5 inches wide. The camera could be swiveled up to 35° either way, to see inside buildings. Its resolution was, at its best, four feet.

Dimensions of observed buildings and objects was for the science of photogrammetry. Horizon and stellar cameras positioned the camera in space precisely. Its precise height was also determined. And photographs of U.S. installations provided scaling data. [Brugioni, p. 363-373 (2010)]

In 1966, the KH-8, was unofficially introduced. It was a 77" camera. It had 6" resolution. The KH-8 satellite weighed 50,000 lbs. or 30,000 lbs. and was 24' or 50' long, and 10' in diameter; electronic transmission. [Richelson (1984)] [AD-A606621 (1973)]

In 1971, there was the fourth generation of spy satellite, the KH-9, or Hexagon, the first "Big Bird". In contrast to the KH-8, it was designed for broad area searches, with a somewhat lower resolution of between one and two feet. Allegedly built by TRW, it was claimed to be 64' long, and weighed 30,000 lbs. Equipped with two cameras with 60" lenses for stereoscopic images, it was a camera mapping satellite.

A silver salt photographic film emulsion is not a very efficient technology. Only about one in twenty photons striking photographic film triggers the chemical reaction with the silver salts that record the photographic image. Thus 95% of the incident light is wasted. In 1970, scientists at Bell Labs developed a silicon chip, the Charged-Coupled Device or CCD, that recorded 75% of the incident photons, and increase of fifteen times over regular silver film. [Kaufmann, p. 111-112 (1985)]

The KH-11 Crystal, developed by TRW, was launched in December 1976, developed jointly by the Air Force and the CIA, and returned PHOTINT from its CCD silicon chip array by digital radio rather than the previous film cartridge system, and its pain-in-the-ass retrieval system. The KH-11 was a "real time" system, allowing for instantaneous observation of imagery, and analysis. It has a two year lifespan, four times longer than the KH-9. It was 64' long, 10' in diameter, and weighed about 30,000 lbs., with a primary mirror 7' 8" in diameter. [Richelson, p. 200 (2001)] [Richelson, p. 55 (2009)] Visible, IR and UV sensor-equipped. William Kampiles, sold its technical manual to the Russians for a paltry $3,000, as revealed in his November 1978 trial.

In 1985, the Space Shuttle launched an ELINT (telephone and radio communications, military radar, and missile radio telemetry) spy satellite, Aquacade, into a geosynchronous orbit from the Space Shuttle. [AD-A235 080 (1990)]

The KH-12 (launched in 1986) had an estimated 4" (10 cm) resolution and was described by fas.org, as an inverted Hubble telescope. The KH-12 (aka "Improved Crystal") thermal imaging possibly; 14 tons (18 tons in later versions), including 7 tons of fuel. Uses synthetic aperture radar (SAR) to look through clouds; amount of back scatter of microwaves reveals contour of ground. View in pairs for better performance. Was allegedly built by Lockheed, Sunnyvale, California and had a 10' diameter.

Then there are the ONYX series, four of which are in Low Earth Orbit, the latest ONYX 5, launched in 2005. The first ONYX 1 was launched in late 1988. They use SAR radar imagery to see through clouds and at night. It is similar to the Magellan probe launched to map the surface of Venus, through its heavy cloud cover.

New satellite reconnaissance activity included a stealth imagery satellite, code-named "Misty" and first launched in February 1990.

In 1992, a new generation of KH satellites began to be launched. [Paglen, p. 103-109 (2010)]

First there was the CIA's Photographic Interpretation Division, with 13 people in 1953. PID became the National Photographic Interpretation Center (NPIC), then in 1996 NPIC merged with the Defense Mapping Agency to become NIMA, the National Imagery and Mapping Agency. Finally, in 2003, NIMA became the National Geospatial Intelligence Agency, probably employing 9,000 employees. [Paglen, p. 123 (2010)]

The civilian Hubble Space Telescope, provides a good comparison with military spy satellites. High orbit (380 miles above the Earth's surface); less than a minute for a detailed image to be down-linked. Spy satellites use LEO of about 200 miles. Amount of blur, b = (h x L)/d, where d is the diameter of the mirror or lens, h is the height of the satellite, and L is the wavelength of light. Blur calculates at 2.4" which is great, but at LEO speed of 5 miles/s, the "dwell time" of the satellite over the target is only 80 seconds.

Military ground terminal down-link at White Sands, New Mexico, a part of the TDRSS (Tracking and Data Relay Satellite System) received military encrypted signal, which is stripped out before civilian access is allowed on retransmission. The TDRSS is owned and operated by a DoD contractor, the Contel Corp., who inhabit the "Blue Cube" in Sunnyvale, CA controlling five or six Keyhole (KH-9 to KH-11+) satellites , as well as other ELINT assets, at the same time as Hubble.

"gyroscope" derived from the Greek, meaning "to view the turning". Does not change its orientation; it remembers its position -- or orientation -- in space. It detects and measures changes in the orientation of the spacecraft.

Each Hubble gyro weighs only two pounds and floats almost frictionlessly in a mixture of 90% liquid hydrogen and 10% liquid helium. Hubble is a Keyhole-class satellite

26 foot long tube; lined with razor-sharp baffles and fins inside the tube to diminish scattered light, such as from the sun, or scattered light from the Moon or Earth; gyros for pitch, yaw, and roll; Aiming and repositioning controlled by electronically activated "reaction wheels" (also called "momentum wheels"),

Four 2-foot diameter, 100 lb. flywheels that obviate the need for rocket thrusters to rotate in three dimensions, to reposition or "slew". Wheels spin normally at 3k RPM, steadying the Hubble to less than .0007 arc seconds, better than any earth-based telescope, that is subject to earth and man-made vibration. Slow: takes an hour to, for example, rotate completely. Checks and fine reposition after a long slew by stellar navigation: confirms its position relative to the location of three well-known, bright stars: Polaris (the North Star), Vega, and Canopus. This rough pointing routine has an aiming accuracy of about an arc minute.

It weighs 12.3 tons (11,200 kg), is 43.5 feet (13.3 m) long, and 14 feet (4.3 m) in diameter, with its solar panels stowed.

Light is reflected to a 12" (0.3 m) diameter convex secondary mirror, which in turn, reflects the light back through a small two foot diameter hole in the doughnut-shaped primary (main) mirror and on past it to the instrument bay behind it, in the spacecraft's stern.

Focal length is 189 ft. (almost 58 m) reduced by using a Cassegrain design, named after a French cleric who designed it during the 1600's; actually a Cassegrain variant known as a Ritchey-Chretien telescope named after two 20th Century opticians who improved the design (by modifying a 1637 treatise by Rene Descartes) eliminated distortion by using hyperboloids rather than paraboloids. Focal ratio, the ratio of focal length divided by diameter of its main mirror is f/24. [Chaisson (1994)]

Thirteen ton HST, launched in 1990. Perkin-Elmer, in Connecticut renamed in 1989 Hughes-Danbury Optical Systems when bought by Hughes Aircraft Co.

In 1979 they made the 94.5" (2.4 m) diameter primary mirror, with ten billionths of an inch smoothness/ focusing precision. The telescope's resolving power was 0.06 second of arc. Mirror blank actually two 2-inch thick face-plates fused to a 10" thick core of glass ribs arranged in a rectangular intermediate grid pattern (like a honeycomb). The grid gave the faces rigidity, while saving weight, at only 1,800 lbs. Ground while lying on a novel fixture containing 134 counter-weighted titanium levers (called a "bed of nails"). each designed to push upwards with a force equal to the weight of the glass above it, simulating the gravity-free weightlessness of space.

Ground to a smoothness deviation of less than 0.01 microns (μm). Vacuum coated with 0.065 microns thick aluminum, and then finally 0.025 microns of magnesium fluoride, to protect the Al from oxidation, and enhance it UV reflectance. Finished mirror had a total reflectance of 85% at visible wavelengths, and 70% in the UV range, and mounted in a titanium ring on its circumference and with a stiff, light-weight graphite fiber-epoxy backing, coated in aluminized Mylar for heat uniformity.

Much of the tube and baffles are coated with Martin black, a code name for a classified, non-reflective coating made by Martin-Marietta Corp.

Light reflected by a 12" (0.3 m) secondary convex mirror 16' up from primary mirror. Made of another kind of low-expansion specialty glass ceramic known as Zerodur, with an aluminum coating and mag fluoride over-coat. Telescope focus 5 ft. (1.5 m) behind primary mirror.

Primary mirror made of ultra low expansion silica glass, doped with titanium dioxide, with an almost zero coefficient of expansion.

CCDs invented in 1970 by Bell Labs, were light, and had good performance across the light spectrum, but only below UV, where they had a very poor response. "They were excellent detectors, not only capable of accumulating light very precisely for long periods, but able to pin down very accurately the relative location of each photon across the entire two-dimensional image..." The first CCDs had a field of view of less than 0.2 inches square, producing extremely small images. They also had to be kept very cold, -100 °F, "in order to reduce the noise produced from the ambient heat."

Through the mid-1970's CCD technology improved dramatically from arrays of 40 pixels to a side (1,600 total) to 400 to a side (160,000), and less temperature sensitive. In 1976, it was proposed to make a detector out of a mosaic of four 400 x 400 CCDs, to make one CCD of 800 x 800 pixels.

Each CCD measured almost half an inch on a side, and was subdivided into an array of 800 x 800 tiny, square picture elements, called "pixels" for short. The four CCDs together form a full image of more than 2.5 million (1,600 x 1,600) pixels. And with 4 bytes per pixel and 8 bits per byte, a typical image contains about 10 Mb or 80 million bits of information. The planetary camera has a focal ratio of f/30, being able to image an area only 1.1 arc minutes across, and ideally resolves objects to about 0.04 arc seconds.

Normally CCDs detect radiation from 4,000 to 11,000 angstroms, not UV. To overcome the UV shortcoming, the CCD was coated with Coronene, a trade-name for a special organic phosphor, discovered by a literature search. A very old technique for spectroscopy, whereby a photomultiplier [tube] was coated with a material that fluoresced under UV radiation. Cooled to -90 °C (-130 °F) to diminish background noise/increase sensitivity.

10x more sensitive than photographic film, capturing 70% of impinging photons. Earth-bound there is the Very Large Telescope in Chile, on a mountain-top, that is made up of an array of four 8-meter diameter mirrors, equivalent to a single mirror 16 meters in diameter.

Mid-2004 two 2,048 x 4,096 CCDs making one 4,096 x 4,096 CCD imager more than 10x larger than the original CCDs installed on Hubble's first camera.

New computer and laser technologies for ground-based telescopes, called "adaptive optics" that was expected to cancel out the unsteadiness of the earth's atmosphere. By 2005, may exceed the resolution of Hubble's 2.4 m diameter, and even 4 m diameter mirror space optics at visible wavelengths. A 3 m diameter primary mirror-equipped telescope estimated to weigh around 25,000 lbs.

Then there is the problem of "jitter" to overcome, arising from gyroscopic motions. If the image started drifting slightly, should the entire telescope body be moved to compensate, or the secondary mirror? Body motion was simpler.

Shorter wavelengths of UV require more precise optics -- the grinding of mirror to higher tolerances. [Zimmerman (2008)] Reconnaissance satellites don't require the UV ability.

By 1989, the Lacrosse, an 18 ton spy satellite, was able to take imagery through darkness and cloud. Lacrosse-1 in 1988. 1,500 radar pulses a second; 50 meter end-to-end solar panels.

24 NAVSTARs at 11,000 km in medium earth orbit (MEO); 1 meter location; "overhead assets"

The super-secret NRO, the National Reconnaissance Office, was established in 1961, under the DoD. Even its **existence** was classified until 1992. Its HQ is in Chantilly, Virginia. It manages the design and construction of reconnaissance satellites by government contractors, principally Lockheed and TRW.

By 1999, commercial U.S. satellite companies were selling imagery with a 1 m (3') resolution. Researchers have suggested that U.S. intelligence agencies have resolution down to 3 inches.


**A Chronology of Spy Satellite Books**

Klass, Philip "Secret Sentries in Space" (1971).

Perry, Robert. "A History of Satellite Reconnaissance", vol. 1 – 6, Corona, NRO (1973)

Greer, Kenneth "Corona" "Studies in Intelligence" 17:1-37. CIA (Spring 1973).

Synthetic Aperture Radar: "Sci. Am." 247:54-61 (Dec. 1982).

King-Hele, Desmond. "Observing Earth Satellites" (1983); the hobby's bible.

Internet Listserv "SeeSat-L".

"Jane's Defense Weekly" (August 11, 1984, p. 171 - 173); leaked spy satellite picture of Soviet battleship; August 1985 trial of NSA official Samuel Morrison; sentenced to two years in prison.

Adams, John A. "Counting the Weapons'. IEEE Spectrum 23:46-56 (July 1986).

Zorpette, Glenn, "Monitoring the Tests". IEEE Spectrum 23:57-66 (July 1986). CCD's.

Burrows, William E. "Deep Black: Space Espionage and National Security". NY: Random House (1986) [unofficial history of Corona and Keyhole]

"High Resolution Radar" Donald R Wehner, Norwood, Mass: Artech House (1987).

Beardsley,Tim. "Hubble's Legacy", Sci. Am. 262:18-22 (June 1990).

Richelson, Jeffrey. "America's Secret Eyes in Space: The U.S. Keyhole Spy Satellite Program". NY: Harper and Row (1990).

National Reconnaissance Office. "Corona (satellite)", Intellipedia. [From www.BlackVault.com]


Excellent discussion of image resolution and other details of satellite reconnaissance:

See Federation of American Scientists, fas.org, "IMINT 101: Introduction to Imagery Intelligence" fas.org/irp/imint.imint_101.html

Richelson, Jeffrey. "The Future of Space Reconnaissance" Sci. Am. 264:38-44 (Jan. 1991)

Hough, Howard. "Satellite Surveillance". Port Townsend, Washington: Loompanics Unlimited (1991).

Kumar, Muneendra. "World Geodetic System 1984" "Surveying and Land Information Systems" 53:53-56 (1993).

Petersen, Carolyn Collins and Brandt, John C. "Hubble Vision". NY: Cambridge University Press (1995).

Ruffner, Kevin "Corona", CIA (1995).

McDonald, Robert A. "Corona Between the Sun and the Earth: The First NRO Reconnaissance Eye in Space". American Society for Photogrammetry and Remote Sensing (1997).

Day, Dwayne A., et al. "Eye in the Sky: The Story of the Corona Spy Satellites". Washington, D.C.: Smithsonian Institute Press (1998).

Day, Dwayne A. "Mapping the Dark Side of the World, Part 1: The KH-5 Argon Geodetic Satellite" "Spaceflight" 40:264-269 (1998).

"Part 2: Secret Geodetic Programs after Argon" "Spaceflight" 40:303-310 (1998).

Monmonier, Mark. "Spying With Maps". Chicago, Illinois: University of Chicago Press (2002).

**The Vela Hotel Spy Satellite**

The Vela (Vela Hotel) NUDET (nuclear detonation) detection satellite research was begun around 1960. [GACAEC, 75th, p. 13 (1961)] Vela is Spanish for "watchman" (Hotel is the radio mnemonic for the letter "H"; possibly it was VH instead of Keyhole, the KH series of PHOTINT satellites). The purpose of the Vela satellites was to detect NUDETs in outer space, in the atmosphere, and on the ground's surface. The satellites would be place in a circular, at a distance of 115,000 km, beyond the Van Allen Radiation Belt. Each pair were spaced at 180° in orbit. In the end, the Vela satellite was considered a very successful program.

The Vela satellites were deployed in pairs. The first pair of satellites was launched (in the same rocket) in 1963, and the final sixth pair, Vela 11 and 12 were launched in April 1970, and operated well into the 1980s. The spacecraft were built by TRW Systems, the radiation sensors by LANL, and the logics by Sandia Corp.

The original plan called for a constellation of eighteen satellites to provide complete Earth coverage. They weighed about 600 lbs. and had a polyhedral shape (a 20 planar equilateral triangle-faced icosahedron, with 18 of its 20 sides covered with solar cells) with a diameter of 50". They orbited the earth in a circular geostationary orbit 67,000 miles above the earth, and received signal from earth at 374 MHz, and transmitted signals to earth at a frequency of 400 MHz at a signal strength of 4 watts. They originally could receive 64 different radio commands from earth, and later 128 commands. The memory storage unit was 30,000 bits of payload data. For reliability/redundancy purposes, many of the Vela's systems were duplicated. [AD483 114 (1965)]

They worked partly based on the double light pulse detected by a bhang meter; a "pulse consists of a rapid 100 μs rise in light intensity, followed by a minimum after 10 ms, and a second maximum after 100 msec. Such signals with two intensity maxima, are typical of nuclear explosions in the atmosphere." [Goldblat, p. 239 (1988)]

Work began at Livermore in FY1960, funded by ARPA to determine the background x-ray, gamma ray, neutron, and other radiation in space, including that due to the earth's Van Allen radiation belts, and in the outer atmosphere. ["Livermore's Director Files", "11-4.1 Program Letters" folder (ca. 1961)]

The first Vela satellite pair was deployed in support of the Limited Test Ban Treaty (LTBT) of 1963. The last and sixth pair was launched in 1970. The satellites had an operating life span of seven years, before starting to degrade, but in fact worked for years.

The Vela satellite had a pair of optical sensors, called "bhang meters", neutron detectors, eight x-ray detectors, four gamma ray burst detectors, an extreme ultraviolet detectors, two Geiger counters, and an EMP sensor. Large cones of beryllium composed part of the Vela Hotel satellite. They used the beryllium's x-ray transparency and visible light opacity in their functioning. [AD331 804 (1962)] The bhang meters detected the unique, very bright double flash of a nuclear explosion above or at the surface of the earth. The double flash lasts one second for a 19 kt atmospheric blast, and has a first peak at about .3 ms, followed by a valley at 12 ms, followed by another much longer peak pulse at 130 ms. This is double pulse is a unique signature of a nuclear explosion. Even a lightning bolt – which has similarities in rare occurrences – does not reach the characteristics of the nuclear double flash. [LASL-79-84 (1979)]

The satellites had 12 cubic shaped x-ray detectors at each of its apexes, with gamma ray and neutron detectors inside. The bhang meters were the principal detectors. A Vela signal detector is found in U.S.P. #4,132,947 (filed in 1967; declassified in 1979)

In July 1983, the Vela functionality was given over to the GPS navigation/location system which had the combined function of NUDET. The GPS system has an eighteen satellite constellation, an orbiting altitude of 20,200 km (within the outer Van Allen Belt) and 55 degree orbits. [Reed (1990)]

The was also the Vela Uniform program for earthbound detection, such as seismic data, as well as the first Nudet program, AFOAT, for airborne collection of drifting radioactive clouds from above-ground testing. It was the AFOAT program that detected the first Russian A-bomb test in 1949.

There is other fallout testing, as in the 1979 South Africa/Israel suspected nuclear test "incident"/ controversy, where elevated levels of iodine-131 were detected in western Australian sheep by researchers.

There is a global network of detection instrumentation, including sound wave detection, seismic wave detection, and ocean hydro-acoustic pulses traversing the world's oceans. And radio telescopes, such as in Arecibo, Puerto Rico, detect ionospheric disturbances – a "ripple" and the direction it is received from. [DeLucas (1997)]

The constellation of geostationary NAVSTAR GPS satellites will also be nuclear-detection capable. A constellation of 18 satellites, controlled by ground stations in NORAD in Cheyenne, Colorado (now renamed SPACECOM), Diego Garcia, Ascension Island, Hawaii, Guam, and the Philippines. It can also be used for ICBM targeting guidance.

**The DSP Early Warning Spy Satellite Constellation**

"Where's the infrared data?  The bandwidth?  The orbiting parameters?  **The frequencies**?  The **physical** frequencies?"

-- Alex, Soviet KGB apparatchik
"The Falcon and the Snowman" (1984)

The Infrared (IR) ICBM launch satellite detection system is a direct extension and outgrowth of the technology initially developed for the Sidewinder (AIM-9A) air-to-air missile in 1956.  Joe Knopow of Lockheed is credited in 1956 - 1957, with the concept of satellite-borne infrared sensing of missile plumes of ICBMs and SLBMs from launch through boost phase to give the U.S. a full 25 - 30 minute warning of a Soviet nuclear attack.

The concept however, was a bit older. It was first enunciated in an earlier RAND report [RM-1572 (1955)], where IR detectors were proposed in planes circling the Russian border, missiles and satellites still being in the future.  The RAND report was further amplified by a NRL (Naval Research Laboratory) report in 1957, giving a pretty complete technological analysis of the concept, based upon the Sidewinder IR photocell technology, the PbS (lead sulfide) IR detector, captured from the Germans at the end of WW2.

In 1958, the development of the WS-117L (Weapon System 117, Lockheed) was authorized by the DoD. The MIDAS (Missile Defense Alarm System) (November 1958 - 1966) was the first attempt at an infrared early warning ICBM launch detection system, to give a thirty minute warning of a Soviet first-strike attack. It used the W-37 infrared sensor, a first generation PbS (lead sulfide) semiconductor built by Aerojet-General, and was the **first** spy satellite the U.S. launched, indicating its priority over the more publicly known Corona (KH) series of PHOTINT surveillance satellites.  MIDAS 1 was first launched on May 24, 1960, and TRW was an associate contractor on the project.

In 1963 a MIDAS satellite detected the test launch of both a Minuteman I and Polaris A-1 strategic missile.  Nonetheless, it was phased out and replaced by the DSP Satellite which had a more advanced sensor design and a more robust spacecraft platform.

IMEWS (1970 - 1972) (Integrated Missile Early Warning System)  The satellite weighed 1,804 lbs. with a five year life span.  IMEWS again used an IR sensor to detect ICBM and SLBM launches.

Finally, there was the development of the first highly successful DSP Satellite (Defense Support Program; "Code 647").  It has been described as "very likely the most successful Air Force satellite program in history." [Dorman, p. VI-3 (1995)]  Aerojet developed both the IR sensor, the W71, and the original ground data processing system.  The price of the IR sensor was about $150 million per sensor in 2012 dollars.

The heart of the system lies with the IR wavelength imaging system used to detect the hot exhaust from missiles. Two different IR wavelength detectors are used to avoid laser jamming. IR sensor systems are very expensive, so a 6,000 IR detector row is rotated through the observation window by spinning the satellite. A complete scan takes about 10 seconds and a launch confirmation requires a number of scans, so verification make take a couple of minutes. The next generation DSP will not require a scan time because well over 100,000 IR detectors may be used, but the cost will be considerably higher

After one failure, a constellation of three more satellites were first successfully launched in geostationary (maintaining the same position over the earth by maintaining the same orbital speed as the earth's rotation) orbit above the equator, at a height of 22,400 miles, starting in May 1971, and used IR (infrared) sensors to detect Russian ICBM or SLBM launches, obsoleting and vastly improving on the NORAD

BMEWS radar system (operational since 1964) that straddled across the Arctic from Clear, Alaska to Thule, Greenland, to Flyingdales, England.

It was the most successful system developed, much better than IMEWS or MIDAS, and consisted of a constellation of multiple satellites for wide geographical coverage of the USSR.

The DSP satellite weighed 2,000 lbs., with a sensor package of 700 lbs., and a power consumption of 400 watts generated by solar cell array panels. It used a 12' long, 3.3' in diameter, wide angle Schmidt IR telescope, over its array of 2,048 PbS (lead sulfide) IR detectors (two 32 x 32 arrays of sensors). The contractor: Aerojet for the sensors and TRW, California ("TRX" in the 1984 spy satellite movie, "The Falcon and the Snowman"), for the overall project. The 2,048 IR detectors eventually evolved to 6,336, and a second array was added to cover additional IR frequency bands. It consumes 1,274 watts of power, and weighs 5,200 lbs.

(The fourth generation of DSP satellite increased the IR sensor pixel count from 2,000 to 6,000 with advances in IR detector technology and decreasing cost. They also now weigh 5,000 lbs. and three can provide coverage for the entire world.)

The satellite consists of an IR telescope with a lengthy barrel, the processing electronics and downlink transmitter, and two star sensors mounted orthogonally to the telescope. So the satellite uses a form of stellar INS. The two sensors are there for redundancy, in case one fails.

The satellite weighs 1,133 lbs. is just under 12 feet high and a 70" diameter. The Link 1 down-link has a word length of 6 bits (this **was** the early 1970's!) and a bit rate of 1,024 kbits/s. The down-link was located at Woomera, Australia.

The interior of the telescope barrel is painted black and heavily baffled to minimize the effects of direct sunlight. The barrel pointed at the earth's surface, the image is reflected off a beryllium primary mirror at the base of the barrel to the sensor package, an array of IR detectors. The signal from the IR detectors is conducted through electrical cabling to the sensor electronics which provides IR detector channel identification and time tagging. Signal processing of the IR data includes filtering, multiplexing, analog-to-digital (A/D) conversion, peak detection, and thresholding (to filter out and discard "noise", and other low-level signals).

For the current satellite, approximately 35 million data samples per second must be compressed to about 23 thousand for the down-link maximum after passing through an analog-to-digital converter.

The PbS sensors were tuned to detect 2.69 to 2.95 micron (µm) range IR signals. Evolutions of the PbS sensor included mercury cadmium telluride (HgCdTe -> HCT) and extrinsic silicon (Si:X), where the "X" could be arsenic or a number of other elements and operated in the longer IR wavelengths, 10 - 30 microns. The extrinsic silicon detectors became extremely popular and they appeared in many astronomical instruments. [Dorman (1995)]

The DSP satellite works by funneling the IR signal through the tubular sunshade, passes through a corrector lens, and is focused on the IR detector array by two mirrors of a Cassegrain configuration, for reasons of length reduction.

A total of 23 DSP satellites were eventually launched. [AD-A511 972 (2009)]

The Teal Ruby satellite (1985) utilizes a 11 foot long cylindrical telescoping system using a focal plane array mosaic of CCD IR detectors. Its focal plane array is divided into thirteen filter zones, with each zone having 1,232 x 96 IR CCD chips, with each array being read a zone at a time. The telescope was made of graphite fiber-filled epoxy, and had a f/3.3 20" aperture. The monolithic silicon array cooled to 15 °K by subliming neon, and the rear optical elements cooled to 70 °K. by subliming methane. [AD-A241 725 (1991)]

It was managed by DARPA and the Air Force and was used for real-time detection of strategic and tactical aircraft and ballistic and cruise missiles.

IR CCDs.  The CCD (Charge-Coupled Device) was a new type of semiconductor developed in 1970 at Bell Labs.

The primary mirrors are made of beryllium metal, which is an excellent IR reflector.  In 1970 Perkin-Elmer had built a polished 35 cm diameter, beryllium mirror coated with electroless gold for smoothness.  The IRAS satellite used a Richey-Chretien beryllium 2-mirror f/9.6 telescope with a 60 cm diameter primary beryllium mirror.

And finally there was the Spitzer f/16 Richey-Chretien Cassegrain beryllium mirror IR telescope with an 85 cm diameter primary mirror.  Its detector is a mix of InSb (indium-antimony) pixels, and Si:As Bib semi-conductor arrays for the 3.6 - 4.5 µm and 5.8 - 8 µm (the IR mid-band), respectively.  Its primary mirror is made of CVD (Chemical Vapor Deposition) SiC (silicon carbide) on porous SiC, with beryllium used as a structural support.  It is an impressive 68.5 cm in diameter and weights only about 31 kg.

*   *   *   *

There were soon many types of spy satellites.  The Defense Mapping Agency used Corona PHOTINT for topographic maps and geodetic maps (for telling where lines of latitude and longitude are drawn).
Millimeter wave radar were developed for their excellent cloud and night penetration ability, but a low resolution, low range, and an orbit of only 115 miles, which limited their operational time.

The GEOSAT satellite, orbited by the Navy was believed to collect information on gravitational force patterns and anomalies for ICBM and SLBM accuracy in their polar orbits over the ice-coated Arctic Ocean.

EXPLORER (1958 – 1961):  Particles, fields, radiation mapping, meteorology, energetic particles, and ionospheric and atmospheric gamma ray measurements.

ELINT (1962 – 1968):  radar (air defense) and radio signals, including missile telemetry.  Second generation satellite named FERRET ("Code 711"; contractors:  Lockheed and Sanders).

The third generation was named RHYOLITE (called "RH"; first launched in March 1973; contractors: Hughes and TRW), and was parked in geostationary orbit 22,300 miles in space where it collected both unencrypted HF and VHF missile telemetry and the microwave signal from long distance telephone transmission links.  The Rhyolite, Argus, and Pyramider programs were disclosed in 1975 to the Soviets by Chris Boyce, in the "Falcon and the Snowman" case, named after the best-selling book about the case, and the 1985 movie.

The fourth generation was named ARGUS (for Advanced Rhyolite).  The down-link for the Rhyolite and Argus satellites was in Pine Gap, Australia, identified in the Chris Boyce espionage trial, and causing a brief local stir in Australia, who had sent reporters to the trial.

GEOS (1965 – 1977) (Geodynamic Experimental Ocean Satellite):  precise measurements of the Earth's surface and shape, producing a mathematical description of its surface and gravitational field.  Originally under the EXPLORER program (Explorer 29 and 36).

DMSP (1971 – 1983) (Defense Meteorological Satellite Program):  Weather data, atmospheric temperature, density, and water vapor content from ground level up to 18 miles altitude, and a microwave imager to pierce cloud cover.  Contractor: RCA.

WHITECLOUD (first launched in 1976):  a super-secret ocean surveillance satellite for nuclear submarine detection, allegedly by tracking their radar and communications.  Developed for the Naval Research Laboratory.  Contractor:  Martin Marietta.

DYNAMICS EXPLORER (1981): contractor: RCA.

If the satellite's launch missile goes east, it is probably an electronic intelligence satellite. If it veers off to the north, it's probably going for a PHOTINT polar orbit.

The Space Shuttle "Endeavor" used synthetic aperture radar to map most of the world's terrain contour in a week.

## The SOSUS Line

The U.S. Navy SOSUS (Sound Surveillance System), the undersea hydrophone cable line was a very effective ASW system installed around 1952 for the long-range detection of passing noisy two-propeller Soviet nuclear missile submarines, using low-frequency underwater sound waves. It was designed and built by Bell Telephone Laboratories for the U.S. Navy. [AD-A241725 (1991)] It covered what was called the GIUK Gap (Greenland, Iceland, and U.K). Norway [JCAE (1973)] was also connected to the SOSUS line, but cared to keep their involvement secret (so it was named GIUK rather than GIUKN…).

The GIUK line was a string of seabed passive acoustic hydrophone detection/surveillance devices, and large coils of wire laid on the seabed to monitor variations in the electric field of the ocean. These were installed along the chock-point for Soviet subs traveling from the major Northern submarine base at Murmansk, on the northern Soviet coast, and allowed the tracking of Soviet "boomers" (ballistic missile submarines) heading on patrol for the Atlantic off the American coast, and having to pass the SOSUS array.

The SOSUS line was begun based on Maurice Ewing's discovery of an underwater sound channel, underneath the thermocline, the bottom of the ocean's upper surface layer of warmer water. The thermocline reflected sounds emanating in the layers above and below it, making complete sound detection to the bottom impossible with a single detector.

In 1954, three additional SOSUS arrays were added to the Atlantic and six in the Pacific Ocean with newer technology. By the late 1960s, a total of over 1,000 deep seabed hydrophones, linked by 30,000 miles (48,300 km) of undersea cable were used in the world-wide undersea network.

The hydrophones were so sensitive that they picked up from a thousand miles away the "crunching" sound of the imploding pressure hull of the 1968 U.S. Scorpion submarine loss/disaster soon after it passed the "crush depth", killing all aboard.

In the days of the noisiest Soviet subs, they could be detected at ranges of several hundred kilometers. In 1980, with much quieter subs, SOSUS could locate it within a circle of 30 km diameter (18.6 miles). Other figures give a value of 10 square miles (26 sq. km) in the location of a disasterously sunken Russian Golf-2 diesel sub.

In declassified patent U.S.P. #3,982,222, was disclosed a primitive form of SOSUS. In 1990, the 1969 U.S.P. #4,965,778 was declassified, giving details of the SOSUS line. It consisted of a cost-effective flexible cord, insulated with RTV silicone rubber, and wrapped within with a toroidal nickel strip winding, annealed to 900 °C, surrounding two sets of wires.

SOSUS, P-3 Orion ASW aircraft, and aircraft carrier groups provide coverage including most of the Northern Hemisphere of both the Atlantic and Pacific Oceans, an array of detectors down both the Eastern and Western seaboards of the U.S., and the Mediterranean. [Schwarz, p. 264 - 265, 303 - 304 (1998)] [Kennedy (1998)] [Arkin, p. 72 - 73,115 (1985)]

## Appendix A:     The History of the U.S. Nuclear Arsenal

| Warhead | Years in Stockpile | Design Lab/Warhead Type |
|---------|-------------------|------------------------|
| Fat Man | 1945-1949 | LANL/ fission implosion bomb |
| Little Boy | 1945-1948 | LANL/ fission gun assembly bomb |
| Mk 3 | 1945-1951 | LANL/ fission implosion bomb |
| Mk 4 | 1948-1953 | LANL/ fission implosion bomb |
| Mk 5 | 1954-1963 | LANL/ fission implosion bomb |
| Mk 6 | 1951-1962 | LANL fission implosion bomb |
| Mk 7 | 1952-1967 | LANL/ fission implosion bomb |
| Mk 8 | 1952-1956 | LANL/ fission gun assembly bomb |
| Mk 9 | 1952-1957 | LANL/ fission gun assembly artillery shell |
| Mk 11 | 1956-1960 | LANL/ fission gun assembly artillery shell |
| Mk 12 | 1954-1963 | LANL/ fission implosion bomb – tactical |
| Mk 14 | 1954 | LANL/ TN (thermonuclear) bomb |
| Mk 15 | 1955-1965 | LANL/ TN bomb |
| Mk 17 | 1954-1957 | LANL/ TN bomb |
| Mk 18 | 1953-1956 | LANL/ fission implosion bomb |
| Mk 19 | 1956-1963 | LANL/ fission gun assembly artillery shell |
| Mk 21 | 1955-1957 | LANL/ TN bomb |
| Mk 23 | 1957-1959 | LANL/ fission gun assembly artillery shell |
| Mk 24 | 1954-1957 | LANL/ TN bomb |
| W-25 | 1957-1985 | LANL/ fission implosion – tactical; Genie air-to-air missile |
| Mk 27 | 1958-1965 | LLNL/ TN bomb |
| Mk 28 | 1958-1991 | LANL/ TN bomb |
| W-30 | 1959-1978 | LANL/ boosted fission implosion - tactical; TALOS Naval missile |
| W-31 | 1953-1989 | LANL/ boosted fission implosion |
| W-33 | 1956-1992 | LANL/ fission Oy gun assembly 8" artillery shell |
| W-34 | 1958-1977 | LANL/ boosted fission implosion |
| Mk 36 | 1956-1962 | LANL/ TN bomb |
| W-38 | 1961-1965 | LLNL/ TN – missile warhead |
| Mk 39 | 1957-1966 | LANL/ TN bomb |
| W-40 | 1959-1972 | LANL/ boosted fission implosion |
| Mk 41 | 1960-1976 | LLNL/ TN bomb |
| Mk 43 | 1961-1991 | LANL/ TN bomb |
| Mk 44 | 1961-1989 | LANL/ boosted fission implosion - tactical; ASROC |
| Mk 45 | 1962-1988 | LLNL/ boosted fission implosion - tactical |
| W-47 | 1960-1975 | LLNL/ TN – Polaris A-1/A-2 SLBM |
| W-48 | 1962-1982 | LLNL/ cylindrical Pu fission implosion 6" artillery shell |
| W-49 | 1958-1975 | LANL/TN - Thor/Jupiter IRBM |
| W-50 | 1963-1991 | LANL/TN - Pershing I tactical missile warhead |
| W-52 | 1962-1977 | LANL/TN – Sergeant SRBM |
| Mk 53 | 1962-1987 | LANL/TN - Titan II ICBM warhead; bomb |
| Mk 54 | 1961-1989 | LANL/fission implosion – tactical; Davy Crockett surface-to-surface |
| W-55 | 1964-1990 | LLNL/U233 fission warhead – tactical; SUBROC missile |
| W-56 | 1963-1993 | LLNL/TN – MM II ICBM |
| Mk 57 | 1963-1993 | LANL/TN bomb - tactical |
| W-58 | 1964-1982 | LLNL/TN - Polaris A-3 MRV SLBM |
| W-59 | 1962-1970 | LANL/TN – MM I ICBM |
| Mk 61 | 1969-1995 | LANL/TN bomb - tactical/strategic |
| W-62 | 1970-1995 | LLNL/TN – MM III ICBM |
| W-66 | 1974-1986 | LANL/TN - Sprint ABM ERW missile warhead |
| W-68 | 1970-1993 | LLNL/TN - Poseidon SLBM warhead |

| | | |
|---|---|---|
| W-69 | 1972-1994 | LANL/TN – SRAM tactical missile warhead |
| W-70 | 1974-1992 | LLNL/TN – Lance tactical missile warhead |
| W-71 | 1975-1992 | LLNL/TN – Spartan ABM missile warhead |
| W-72 | 1971-1978 | LANL/TN – Walleye tactical glide bomb |
| W-76 | 1978-1995* | LANL/TN – Trident I SLBM warhead |
| W-78 | 1979-1995* | LANL/TN – MM III ICBM warhead |
| W-79Mod3 | 1986-1992 | LLNL/TN – ERW 8" artillery shell |
| W80 | 1982-1995* | LANL/TN – ALCM Tomahawk CM |
| Mk 83 | 1983-1995* | LLNL/TN – Strategic Bomb |
| W84 | 1983-1995* | LLNL/TN – GLCM Cruise Missile strategic warhead |
| W85 | 1983-1991 | LANL/TN – Pershing II strategic missile |
| W87 | 1986-1995* | LLNL/TN – MX/Peacekeeper ICBM warhead |
| W88 | 1988-1995* | LANL/TN – Trident II SLBM warhead |

\* Currently in arsenal

[compiled from Francis (1995)]

**Appendix B:**   **Further Details of the U.S. Nuclear Arsenal**

[Cochran, Vol. 2 (1987)]

| Weapon | Years Manu-fractured | Retirement | # Manu-factured | Comments |
|---|---|---|---|---|
| **Pure Fission Implosion weapons** | | | | |
| Trinity | 1945 | | 1 | Test of Nagasaki weapon |
| Fat Man | 1945 | | 1 | 54" dia.; used on Nagasaki |
| Mk 3 | 4/47-4/49 | late 1950 | 135 | Fat Man, Model 1561; 3 mods |
| Mk 4 | 3/49-5/51 | 7/52-5/53 | 175 | assembly line Fat Man |
| Mk 5 | 5/1952 | last ret. 1/63 | 140 (all mods) | 45" dia. aerial bomb |
| W5 | 1954-1955 | 7/61-1/63 | 100 | Regulus I, Matador CM. |
| Mk 6 | 7/51-early 1955 | last ret. '62 | 1,100(all mods) | 54" dia. improved Mk4; 60 detonator/lenses |
| Mk 7 | 7/52-2/63 | last ret. '67 | 1,700-1,800 | 20-60 kt 30" dia. split-levitated core aerial bomb |
| W7 | 12/53 | ret. 1960's | 2,000 | Honest John, Corporal, Boar, Betty warhead, ADM |
| Mk 12 | 1954-1957 | | 250 | 14 kt 22" dia. tactical fighter aerial bomb |
| Mk 18 | 1953-1955 | | 90 (all 2 mods) | hollow core, 550 kt Super Oralloy Bomb (S.O.B.) |
| W-25 | 5/1957-5/1960 | 8/61-1965Mod0 | **2,321** | Genie Air-to-air Missile |
| W-48 | 10/63-3/68 | | 925 Mod1 | 6" diameter artillery shell  Pu cylindrical implosion; 80 tons yield; 684 total artillery pieces |
| Mk 54 | 4/61-2/65 | 7/67-1971 | 1,660 | 20 ton yield; Davy Crockett recoilless tactical shell; 50 lbs.; 10.9" dia.; Falcon Air-to-Air missile, and SADM |
| **Gun assembly fission** | | | | |
| Little Boy | 1945 | | 1 | Hiroshima |
| Mk 8 | 1951-1953 | | 20 | "LC" or "Elsie" for light case; 2 mods |
| Mk 9 | 1952-1953 | | 80 | 11"/280 mm dia. 15 kt artillery shell |
| Mk 11 | 1956-1957 | | 40 | 14" dia.  bomb |
| Mk 19 | 1955-? | | 80 | 11"/280 mm dia. artillery shell |
| Mk 23 | 1956-? | | 50 | 16" dia. Naval artillery shell |
| W-33 | 1957-1965 | | **540** | 8" dia. artillery shell; replaced by W79; 47 kg Oy, Be tamped; 1 kt yield |
| **Boosted fission implosion weapons** | | | | |
| W-30 | 2/59-1/65 | | 600 | 22" dia. boosted Oralloy; TALOS Navy SAM, TADM |
| W-31 | 10/58-12/61 | | 4,450 | 30" dia. boosted, hollow core oralloy; Y3+Y4 Honest John SRBM; Y1+Y2 Nike-Hercules SAM; ADM, the W31 replacing the Mk 7 |
| W-34 | 6/58-12/62 | | 3,200 | 17" dia., boosted; Lulu depth charge; Astor torpedo, Hotpoint tactical bomb; primary for 1.85 MT Mk27. |
| W-40 | 9/59-5/62 | | 750 | **383** Bomarc SAM, Lacrosse SSM |
| W-44 | 5/61-3/68 | | 575 | ASROC Ship-launched ASW missile; 10 kt yield; 12,000 launchers installed on ships; nuclear and non-nuclear ASROCs |
| Mk 45 | 9/61-6/66 | | 1,650 | Little John SRBM, Bullpup B AAM, ship Terrier SAM missiles; |

|          |                 | MADM |                                                                 |
|----------|-----------------|------|-----------------------------------------------------------------|
| W-55     | 1/64-3/68       | **224** | SUBROC submarine-launched ASW missile                        |
|          | 3/70-4/74       |      |                                                                 |

## Thermonuclear Weapons

### (Aerially-dropped bombs)

| Mk 14     | 1954-1954          | 5    | Emergency Capability TN bomb |
|-----------|--------------------|------|------------------------------|
| Mk 15     | 4/55-2/57          | 400  | 1,200 produced (all models) 34.5" dia. 1.7 MT bomb |
| TX-16     | 1/54-4/54          | 5    | cryogenic TN bomb; version of Ivy Mike |
| Mk 17     | 7/54-11/55         | 200  | first production dry LiD TN weapon |
| Mk 21     | 1955-1956          | 275  | 275 (all 3 mods) |
| Mk 24     | 7/54-11/55         | 105  | same dimensions as Mk17 |
| Mk 27     | 1958-1959          | 20   | Navy aerial bomb; Regulus I CM |
| Mk 28     | 1958-1966          | 1,200 | 365 kt/1.1MT/1.44 MT B-52 bomb; Hound Dog and Mace CMs |
| Mk 36     | 4/56-6/58          | 940(all 5 mods) | highest yield warhead produced |
| Mk 39     | 2/57-3/59          | 700(all 3 mods) | 1957-1966;3.75 MT high-yield variant of Mk15; Redstone SRBM, Snark CM |
| Mk 41     | 9/60-6/62          | 500  | Lower yield, lighter replacement for Mk36; ret. 11/63 |
| Mk 43     | 1961-1965          | 1000(all 5 mods) | 880 kt yield; 18" dia. |
| Mk 53     | 8/62-6/65          | 350  | |
| Mk 57     | 1/63-5/67          | 3100(all 6 mods) | there were 825 in 1989; multipurpose aerial bomb; tactical; ASW |
| Mk 61*    | 10/66-early 90s    | 3,150 | 1,350 In service in 1995; originally 7 Mods, older mods retired; some modes used Oy, some U238; uses IHE; 13.5" diameter |
| B61Mod3   | 1979-1990          |      | |
| B61Mod4   | 1979-1990          |      | |
| B61Mod7   | 1985-1990          |      | |
| B61Mod10  | 1990-1991          |      | used some W-85 parts; probably primary |
| Mk 83     | 1983-1990          | 738  | uses IHE and FRP |

## Thermonuclear Weapons

### (Mostly Missile and Cruise Missile Warheads)

| W-38      | 1961-1963          | 180  | Atlas, Titan I ICBM warheads |
|-----------|--------------------|------|------------------------------|
| W-47      | 1960-1964          | 300  | 18" dia x 47" length; Polaris A-1 had 400 kt and Polaris A-2 had 1.2 MT W47-Y2 |
| W-49      | 1958-1964          |      | Thor/Jupiter IRBM; Atlas/Titan I; 1.44 MT variant of also 20" dia. Mk28 |
| W-50      | 1963-1965          | **750** | Pershing I + IA SRBM; Nike-Hercules ABM; W-50 weight 410 lbs. based on RW Huron; 15.3" dia x 43" long; 800 lbs. |
| W-52      | 1962-1966          | 300  | Sergeant SRBM; 13-14" dia.; 225 kt |
| W-53      | 1962-1965          | 54   | Titan II ICBM warheads; 8.9 MT yield; 37" dia. x 103" long; 6,200 lbs. (cf. Mk 53 aerial bomb) |
| W-56      | 1963-1969          | 455  | Minuteman II; Mod4's; based on W-47 |
| W-58      | 1964-1967          | 1,400 | Polaris A-3 MRV; precursor to MIRV; based on HT1, Olive Event; 202 kt, 219 lbs. |
| W-59      | 1962-1963          | 150  | Minuteman I; Mk 5 RV |
| W-62*     | 1970-1976          | **1,650** | Replaced in 1979 by W-78, with double the yield |
| W-66      | 1974-1975          | 30   | Sprint ABM warhead; first ERW (neutron bomb) |

| | | | |
|---|---|---|---|
| W-68 | 1970-1975 | **5,250** | In 1979, HE replaced from LX-09 to -10 |
| W-69 | 1971-1976 | **1,451** | SRAM; Air-to-Ground missile; derived from Mk 61 |
| W-70 | 1974-1992 | 1280 | Lance; 380 Mod3's (ERW version) produced |
| W-71 | 1974-1975 | 30 | Spartan ABM;High fusion TN; 4 MT |
| W-72 | 1970-1972 | 300 | Walleye; yield-enhanced version of Mk 54 |
| W-76 | 1978-1987 | **4,560** | Mk 4 RV; 8-14 warheads loaded |
| W-78* | 1979-1982 | **1,650** | MM III upgrade; design derived from W-50, with lighter primary |
| W-79 | 1981-1986 | 550 (325 ER) | Mod 3 is the ERW version; others fission only |
| W80-0* | 1981-1990 | 370 | Tomahawk SLCM |
| W80-1* | 1981-1990 | 1,400 | ALCM |
| W84* | 1983-1988 | 500 | GLCM; uses IHE; design derived from Mk 61 Mod3/4 |
| W85 | 1983-1986 | **276** | design derived from Mk 61Mod3/4 |
| W87* | 1986-1989 | 525 | uses IHE and FRP; Mk 21 RV |
| W88* | 1988-1989 | 675 | < 368 kg weight; Oy secondary; Mk 5 RV |

* Part of post-Cold War "enduring stockpile", with all other stockpile devices retired

[Numbers in bold verified by DoE documents]

[compiled from Sublette (1997)] [KCD-40391-ENG. "Missile Warhead Cross Reference". (January 21, 1965)]  [SAND95-2751 (1996) "Stockpile Surveillance:  Past and Present" Johnson, Kent, et al.] [Gardiner, p. 146 (1993)]  [Francis, p. 104 (1995)]

# Appendix D: Early Stockpile Operational Dates

## Early Pre-H-Bomb Mostly Pure Fission Atomic Bombs

| Model | Production Date | Retirement |
|---|---|---|
| Mk 3 Model 1561 Production "Fat Man" | 4/47-4/49; | all 135 retired late 1950; 25 kt |
| Mk 4 composite core | 3/49-5/51; | nominal 40 kt; based on 1948 Sandstone X-ray event; retired 7/52-5/53; 550 produced (all 3 mods) |
| Mk 5 comp.core | 5/52-? | all retired by 1/63; 140 produced (all 4 mods); 100 W-5 produced regulus/matador from Mk5 conversion |
| Mk 6 | 7/51-12/53 | 80 kt; levitated; based on 1951 Greenhouse Dog event; Composite Core; last retired '62; 1,100 produced (all 7 mods) |
| Mk 7 | 7/52-2/63 | 30/60 kt; 10 mods; 1,700-1,800 produced |
| Mk 12 | 12/54-2/57 | 14 kt; retired 7/58-7/62; 250 produced, all 2 mods |
| Mk 18 | 54-2/55 | 550 kt; Retired 1/56-3/56; 90 produced, all 2 mods |
| W-25 | 56-5/60 | Ret. 1985; 1,800 produced (all two mods) [Dorman (1995)] |
| W-40 | 9/59-5/62 | 10 kt; Ret. 10/63-1964; boosted fission warhead |

## Early H-Bombs (1954-1958)

| Weapon | Manufacture Date | Number Manufactured |
|---|---|---|
| Mk 14 | 2/54-10/54 | 5 produced; recycled into Mk 17's; 96% Li6 |
| Mk 15 | 4/55-2/57 | Retired 8/61-4/65; 1,200 produced, all 3 mods |
| TX-16 | 1954 | cryogenic Ivy Mike-type device; 5 produced |
| EC-17 | 4/54-10/54 | 5 produced; natural Li |
| Mk 17 | 7/54-11/55 | Retired 1/57-8/57; 200 produced, all 3 mods |
| Mk 21 | 12/55-7/56 | Retired 6/57-1/58; 275 produced, all 2 mods; retired by conversion to Mk 36Y1Mod1 |
| EC24 | 4/54-10/54 | 10 produced; enriched Li6 |
| Mk 24 | 7/54-11/55 | Ret. 9/56-10/56; 105 produced, all 2 mods |
| Mk 36 | 4/56-6/58 | Retired 8/61-1/62; 940 produced, 2 mods |
| Mk 39 | 2/57-'66 | |

| | | | |
|---|---|---|---|
| Mk 41 | | 9/60-6/62 | 16 MT; Ret. 11/63-7/76; 500 produced |
| Mk 53 | | 8/62-6/65 | 8.9 MT; Ret. 7/67; 350 produced, 4 mods |
| Mk 27 | | 11/58-6/59 | 1.85 MT;  Ret. 11/62-7/65 |
| Mk 28 | | 1/58-5/66 | 1961 early mods ret.  Ret. 9/91; 4,500 produced (all mods) |
| Mk 49 | | 9/58-1964 | 1.44 MT |
| W-50 | | '63-'65 | Ret. '73; 750 produced (I + IA); 280 Pershing I's & 470 Pershing IA's produced; 60/200/400 kt |
| Mk 53/W53 | | 8/62-6/65 | 350 produced; Y1 & Y2; Y2 retired; 54 Titan II warheads |
| W-55 | | 1/64-3/68 & 3/70-4/74 | **224** produced |
| W-56 | | 3/63-5/69 | 1,000 produced; early mods ret. 9/66; mod4 ret. 1991-1993 |
| W-58 | | 3/64-6/67 | 1,400 produced; ret. 9/68-4/82 |
| W-59 | | 6/62-7/63 | 150 produced; ret. 12/64-6/69 |
| Mk 61 | Strategic/ Tactical Bomb | 10/66-early '90's | 1,350 In Service |
| W-62 | MM III ICBM | 3/70-6/76 | **1,725** produced |
| W-68 | Poseidon | 6/70-6/75 | **5,250** produced; ret. 9/77-1991; largest production run of any warhead; 31 subs with 16 SLBMs with an average of 10 MIRVs per missile. |
| W76 | Trident I | FY77-FY84 | approx. **4,560** produced |
| W78 | MM III ICBM | 8/79-10/82 | 1083 produced |
| W80-0 | SLCM | 12/83-9/90 | 367 produced |
| W80-1 | ALCM | 01/81-9/90 | 1750 produced |
| B83 | Strategic Bomb | 6/83-1991 | 738 produced and In Service |
| W84 | GLCM | 9/83-1/88 | 500 produced; 425 In Service |
| W85 | Pershing II | 2/83-7/86 | **276** produced; Ret. '88-3/91 |
| W87 | Peacekeeper/ MX ICBM | 7/86-12/88 | **525** produced |
| W88 | Trident II | 9/88-11/89 | 675 In Service |

[W76/W78/W80/B83/W84/W87:  Calculates to the use of 10 kg Beryllium metal per weapon.]
[Sublette (2006)]  [SAND97-8017 (1998)]

**Appendix I:** **Warhead Weight Calculations**

W-62 & W78 Minuteman III RV only weight
=================================

Mk 12 RV weight: approx. 2.25 g/cm3 graphite; 0.5" thick; RV dimensions: 17.25" dia x 60" high

Base: $\pi \times r^2 \times .5 \times 2.54 \times 2.25$ grams = 4.3 kg

Side: tan (half-angle) = 8.63/59.9 = .144; half-angle= 8.2°

$\qquad$ sin 8.2° = 0.146 ; 0.854
$\qquad$ $2 \times \pi \times (2.54 \times 8.63) \times 0.5 \times 2.54 \times [(60 \times 2.54)/(\sin 8.2°)] \times 2.25$ grams
$\qquad$ = 69.5 kg

**Total RV only weight: 69.5 + 4.3 = ~75 kg = 165 lbs.**

[SAND94-0335 (1995)]


W87 MX warhead
==============

RV + W87 = 210 kg = 465 lbs. RV: 21.8" x 68.9" high; half-angle 8.2°

Sides: $2 \times pi \times (2.54 \times 10.9) \times 0.5 \times 2.54 \times [(68.9 \times 2.54)/\sin 8.2°] \times 2.25$ grams
$\qquad$ sin 8.2° = 0.146; 0.854
$\qquad$ RV sides weight = 102 kg

Base wt.= $pi \times r2 \times .5 \times 2.54 \times 2.25$ grams = 7 kg

**RV total weight (not including LiH anti-ABM shielding) = 110 kg.**
**Therefore, Weight of W87 < 210 – 110 kg < 100 kg**


W-68 Poseidon Warhead
====================

[U.S.P. #4,577,812 (1986) Platus]:

W-68 dimensions: 12" dia x 74.4" high, half-angle 4.25°, 74.4/0.926 = 80.3"
(warhead + RV) = 150 lbs.

RV weight: side = $2 \times pi \times (2.54 \times 6) \times .5 \times 2.54 \times 80.35 \times 2.54 \times 2.25$ grams = 55.8 kg = 123 lbs
Therefore weight of W-68 = 150 – 123 = 27 lbs. (?)
-------------------------------------------------------------------------------------------------------

Spartan W-71 4.4 MT warhead hard x-rays? 60 keV.

4.5x yield nat LiD -> 96% Li6D
Castle Romeo dia. of 60" / five feet and yield of 11 MT ; natural Lithium
Diameter of 60", yield of W-71 is (4.5) 11 = 50 MT
60 x 0.8 x 0.8 x 0.8 x 9 = 28" dia. for 4.4 MT

Wt. of nuclear debris.  4 MT energy;  4000 kt / 64 kt/kg = 63 kg Li6D  diameter?
63,000 = 0.88 x 1.33 x 3.14 x $r^3$; r = 25 cm; therefore diameter secondary ~= 20"

## 15 mile radius 3 MT; hardened Russian RV:

3 MT = $3 \times 10^{15}$ cal
surface area 15 miles radius = $4\pi[(15/0.625) \times 10^5]^2 = 7.24 \times 10^{13}$ cm2
41 cal/cm2

W71  4 MT Spartan radius of effect:  4,000/64= about 63 kg Li6D; r = 25 cm = 10"; 20" diameter.
1 kt = $10^{12}$ cal
4 MT = $4 \times 10^{15}$ cal
surface area 50 km radius = $4\pi(5 \times 10^5)^2 = 3 \times 10^{12}$ cm2
4 MT = $1.3 \times 10^3$ cal/cm2 at 50 km radius; Most energy release spread over 4 generation (12 ns);
maximum x-ray heat radiation $(1.3/4) \times 10^3$ = 325 cal/cm2

[AD355 863 (1963)]  A 200 kt explosion at 90,000 feet, has a lethal radius of greater than 0.3 km.
Therefore 200 x 4 x 4 = 3.2 MT has a lethal radius of 3.6 km for an unhardened Russian RV.

*************************************
AD- A995 074, p. 25 (1979)]

The cross section for the B10 (n, α) Li7 reaction is 4,020 barns for thermal neutrons.  The reaction is
exothermic, with a release of energy of 5.1 MeV (2.8 & 2.3 MeV for the alpha particle, and the Li7,
respectively).  There is 20% B10 in natural boron (with the rest B11).

Pu shielded/surrounded by a 2 cm thick layer of Boron10 has an effective cross section of 1.5 – 2 barns
at from 0.5 to 14 MeV, respectively.

$$\sigma_{eff} = \sigma_f \times e^{-\sigma B \, NB}$$

where  $\sigma_{eff}$ = effective cross section of the Pu shielded by Boron10
    $\sigma_f$ = fission cross section of the Pu
    $\sigma_B$ = cross section of B10 (n,α) Li7
    $N_B$ = number of B10 atoms per cm2 of the surface area of the boron shield

0.7 b cross section of boron at 0.5 MeV;
 ~0.05 b at 14 MeV
Fission cross section Pu at 0.5 MeV = $1.9 \times 10^{-24}$

Boron10:  atomic weight 10, density = 2.35 g/cm3
1 mole = 10 g = $6 \times 10^{23}$ atoms;  2.35 grams = $(2.35/10) \times 6 \times 10^{23}$ atoms in a cm3 = $1.41 \times 10^{23}$
Single layer of boron atoms = $(1.41 \times 10^{23})^{1/3} = 5.2 \times 10^7$ atoms in a cm2

Effective cross section at 0.5 MeV = $1.9 \times 10^{-24} \times e^{-7 \times 10^{-25} \times 5.2 \times 10^7}$ = 0

*********************

| | |
|---|---|
| Ballistic | The trajectory of a projectile, following the end of the propelling force, at which point it is only acted upon by its momentum, gravity and aerodynamic drag, i.e, a non-propulsive free-fall. |
| CEP | Circular Error Probable. Defined as the radius of a circle, centered on the aim point (target), within which 50% of the RV's are expected to impact. Sometimes referred to as the "miss distance". |
| DGZ | Designated Ground Zero. The desired target. A SIOP term. |
| FOBS | Fractional Orbit Bombardment System. A system developed by the Soviets that launched ICBMs from the unusual direction of south to hit the U.S. from any direction, and avoid the BMEWS, early warning system radar which expected a missile/bomber attack from over the North Pole. |
| Ground Zero | The point on the ground directly below the detonation of an nuclear weapon in the air (an air-burst). |
| Fairing | The aerodynamically shaped "nose cone" covering of the MIRV warhead, which protects them from atmospheric effects. It is ejected in the exo-atmosphere (after the missile has boosted itself outside the atmosphere). |
| ICBM | A strategic ballistic missile with a range of 3,100 miles (5,000 km) or greater. |
| IRBM | A ballistic missile with a range of between 1500 to 3000 nm. A European-based strategic missile |
| Launch Vehicle | A rocket for sub-orbital (missile), orbital (satellites) or escape velocity (moon launch, mars exploration, etc.). |
| MAD | Mutual Assured Destruction; the basis of the idea of nuclear deterrence, that nuclear war would cause unacceptable damage to both sides, and the resulting, uneasy, nuclear-armed peace. |
| micron | $1 \times 10^{-6}$ m; a millionth of a meter. |
| mil | .001"; a thousandth of an inch. |
| MIRV | Multiple Independently Targeted Reentry Vehicle. A missile equipped with multiple warheads which can strike differently located targets within its comparatively large PBV target scope ("footprint"). An evolution from the earlier MRV. |
| MRBM | A ballistic missile with a range of between 600 and 1500 nm. |
| MRV | Multiple RV. Multiple smaller yield warheads for a single missile, and a single target. |
| nm | Nautical mile = 1.2 statute miles. |
| NIE | National Intelligence Estimate. A Top Secret document that summarizes the position of the collective U.S. intelligence community on issues of concern. |
| Oralloy | "Oak Ridge Alloy"; an old code-name for uranium enriched to 93.5% U235 for use in weapons. |

| | |
|---|---|
| PBV | Post Boost Vehicle, or "Bus". The final liquid-fueled stage of a MIRVed ICBM. The PBV carries the MIRV warheads, Pen Aids, and its own guidance and propulsion systems. |
| Radiation | The emission and propagation of energy through space or a material medium. Only alpha, beta, or neutron particles or x-rays or gamma rays are intended for nuclear detonations or detonation products. |
| SIOP | Single Integrated Operational Plan; the highly classified U.S. nuclear war fighting plan, including targets for nuclear destruction specified to the specific carrier and warhead, and under various scenarios, including Total War. |
| Shock wave | A high pressure wave expanding outwards from its source at a speed higher than the speed of sound in the material it is in. It travels outwards compressing the material in the lower density material in front of it to a high pressure. |
| Specific Impulse | $I_{SP}$. Rocket thrust divided by weight flow of propellant, in units of seconds. Probably the best way of thinking of specific impulse is as a velocity expressed in units of 9.8 m/s2 (acceleration due to gravity). |
| SRBM | Short-Range Ballistic Missile |
| Strategic | Weapons are either tactical (battlefield) or strategic: targeted directly on the enemy country. The post-9/11 "Homeland Security Agency" is actually "Strategic Security". Due to U.S. geography, the only previous strategic attacks or threats to U.S. strategic security were Pearl Harbour, and Soviet ICBMs. |
| Throw-weight | The weight of a missile's payload, including nuclear weapons, RV, navigation and guidance system, and PBCS (bus). In other words, the weight of the rocket not connected with propulsion, like the rocket casing, main engines (excluding the MIRV bus), and fuel. |
| War Reserve | Fully manufactured item, to the appropriate specifications, for use as a weapon during a nuclear war. |
| Yield | The total effective amount of energy released by the explosion of a nuclear weapon. It is usually expressed in terms of equivalent tonnage of TNT required to produce the same explosion. |

**Appendix K:**                                        **Acronyms**

AEC            Atomic Energy Commission; latter renamed ERDA, and then finally the DoE

ABM            Anti-Ballistic Missile

ACM            Advanced Cruise Missile

ACDA           Arms Control and Disarmament Agency; eventually absorbed by the Dept. of State

AF             Air Force

AFOAT          Air Force Operation Atomic Energy

AIFI           Automatic Inflight Insertion

aka            also known as

ALCM           Air-Launched Cruise Missile

ARPA           Advanced Research Projects Agency.  In 1972 changed its name to DARPA

ASROC          Anti-Submarine Rocket

ASW            Anti-Submarine Warfare

AWRE           [U.K.] Atomic Weapons Research Establishment, aka "Aldermaston"

CCD            Charged-Coupled [semi-conductor component] Device

CEP            Circular Error, Probable

CIA            Central Intelligence Agency

CM             Cruise Missile

COCOM          Coordinating Committee (western country committee for deciding on the export/non-export of dual-use proliferant technology)

CTBT           Comprehensive Test Ban Treaty

DARPA          Defense Advanced Research Projects Agency; before 1972 known as ARPA

DoD            [U.S.] Department of Defense

DoE            [U.S.] Department of Energy

DoJ            [U.S] Department of Justice

DTIC           [U.S.] Defense Technical Information Center

ECM            Electronic Counter-Measures

EM             Electromagnetic (Radiation)

EMP            (Nuclear) Electro-magnetic Pulse

| | |
|---|---|
| ER | Enhanced Radiation (Neutron Bomb) |
| ERDA | Energy Research and Development Agency (successor to AEC, predecessor to DoE |
| ERW | Enhanced Radiation Weapon (Neutron Bomb) |
| FOIA | [U.S.] Freedom of Information Act (5 U.S.C. Section 552) |
| FRD | Formerly Restricted Data; classified yields, location, numbers, and other military information about nuclear weapons |
| FRP | Fire Resistant Pit |
| g | Force of gravity |
| GACAEC | General Advisory Committee of the AEC |
| GLCM | Ground-Launched Cruise Missile |
| HE | High Explosive |
| HEU | Highly Enriched Uranium (>20% U235) |
| HOB | Height of Burst |
| ICBM | Inter-Continental Ballistic Missile |
| IFI | (Manual) Inflight (fissile core) Insertion |
| IMU | Inertial reference and velocity Measurement Unit |
| INF | Intermediate Nuclear Forces (Treaty) |
| IOC | Initial Operating Capability |
| IRBM | Intermediate-Range Ballistic Missile |
| IR | Infra-red |
| JCAE | [U.S.] Joint [Congressional] Committee on Atomic Energy |
| JPRS | [U.S.] DoD Joint Publications Research Service (translates foreign language documents) |
| kt | kilotons of nuclear yield;  the nuclear yield in equivalent number of 1,000 tons of TNT |
| LANL | Los Alamos National Laboratory |
| LASL | Los Alamos Scientific Laboratory |
| LiD | the chemical symbol denoting lithium deuteride, the main thermonuclear fusion fuel |
| LLNL | Lawrence Livermore National Laboratory (Livermore, California) |
| LRL | Lawrence Radiation Laboratory (post-UCRL re-naming, and predecessor of final LLNL name) |

| | |
|---|---|
| LTBT | Limited Test Ban Treaty (1963) |
| MAD | Mutual Assured Destruction |
| MaRV | Maneuvering Reentry Vehicle |
| MGS | Missile Guidance System |
| MIRV | Multiple Independently-targeted Reentry Vehicle |
| Mk | Mark |
| MM | Minuteman ICBM |
| MRBM | Medium-Range Ballistic Missile |
| MRV | Multiple Reentry Vehicle; the precursor to the more advanced MIRV |
| MT | Megatons of nuclear yield; the nuclear yield in equivalent number of millions of tons of TNT. |
| MT | Metric Ton (2,000 lbs.); as opposed to the regular 2,200 lb. ton |
| NATO | North Atlantic Treaty Organization |
| nm | Nautical Mile (= 1.151 statute miles = 1,852 meters = 2,025 yards) |
| NGO | Non-Governmental Organization (a private, independent lobbying organization) |
| NPT | Non-Proliferation Treaty |
| NSC | United Nations National Security Council |
| NuDet | Nuclear Detonation |
| NYT | The "New York Times" newspaper |
| OAD | Operational Availability Date |
| Oralloy | the old code word, "Oak Ridge Alloy"; HEU of ~93.5% U235 weapons-grade enrichment |
| Oy | Oralloy |
| PAL | Permissive Action Link; nuclear weapon unauthorized detonation prevention safety device |
| PBCS | Post-Boost Control Sub-System ("Bus") (PBV) |
| PBV | Post-Boost Vehicle ("Bus") (PBCS) |
| PRO | Public Records Office (U.K.) |
| Pu | Plutonium; fissile Plutonium-239 |
| Pu239 | Plutonium-239 |

| | |
|---|---|
| RB | Re-entry Body |
| RBA | Re-entry Body Assembly (RV with release assembly) |
| RD | Restricted Data |
| REB | Re-entry Body |
| RFP | DoE Rocky Flats Plant, Colorado |
| RV | Re-entry Vehicle |
| S-RD | Secret-Restricted Data |
| SALT | Strategic Arms Limitation Treaty |
| SIOP | Single Integrated Operational Plan |
| SLBM | Submarine-Launched Ballistic Missile |
| SLCM | Sea-Launched Cruise Missile |
| SNM | Special Nuclear Material |
| SRAM | Short-Range Attack Missile |
| SRBM | Short-Range Ballistic Missile |
| SRD | Secret-Restricted Data |
| SRF | [Soviet] Strategic Rocket Forces |
| SSBN | Sub-Surface Ballistic Nuclear (SLBM submarine) |
| SSN | Sub-Surface Nuclear (Counter-Submarine Attack Submarine) |
| START | Strategic Arms Reduction Treaty |
| STP | Standard Temperature and Pressure |
| SUBROC | Submarine Rocket |
| TN | Thermonuclear |
| TNA DEFE | The National Archives (Kew, England) – DEFENCE files |
| TBT | Threshold Test Ban Treaty |
| U235 | Uranium-235 |
| UCRL | University of California Radiation Laboratory (predecessor organization of LLNL) |
| UDMH | Unsymmetrical dimethyl-hydrazine |
| USC | United States Code |
| USSR | Union of Soviet Socialist Republics |

| | |
|---|---|
| W | Warhead (Missile) |
| WR | War Reserve |
| WTO | Warsaw Treaty Organization's Allied Nation(s); the Warsaw Pact |

**Appendix L:**                 **Physical Constants and Physics Equations**


## Conversion Factors

1 kt fission = $10^{12}$ calories = $4.184 \times 10^{12}$ Joules = $2.62 \times 10^{25}$ MeV = $3.97 \times 10^{9}$ BTU = $4.18 \times 10^{19}$ ergs = total fission of approx. 57 g U235 = $1.5 \times 10^{23}$ fissions

1 mole = $6.02 \times 10^{23}$ molecules; 1 mole of any gaseous element = 22.4 liters (at STP)
Absolute zero = 0 °K. = -273.15 °C.

The 4 MT W-71 Spartan ABM warhead gives off 11 cal/cm2 at 50 km radius [SAND91-0285 (1992)], at 3 m: $1.1 \times 10^{30}$ cal/cm2; therefore 1 kt at 3 m: $3 \times 10^{26}$ cal/cm2

Stefan-Boltzmann Constant, $\delta = 5.67 \times 10^{-5}$ ergs/cm$^2$.K$^4$.s = $8.9 \times 10^{-12}$ J/cm$^2$.K$^4$.s
Energy radiated based on temperature = E = $(\delta)$T$^4$; T in degrees Kelvin, E is in ergs/cm$^2$.s

For 4.5 keV: E (in ergs/cm$^2$.s) = $5.7 \times 10^{-5}$ ergs/cm$^2$.K$^4$.s x (4.5 keV x $11.6 \times 10^{6}$ °K/keV)$^4$ = $5.7 \times 10^{-5}$ x $(52.2 \times 10^{6})^4$ = $(4.2 \times 10^{26})$ ergs/cm2.s ; $(4.2 \times 10^{26})$ / $(4.2 \times 10^{7})$ = $10^{19}$ cal /cm2.s = $10^{10}$ cal/cm2.ns;
For 1 keV: $(11.6 \times 10^{6})^4$ = $(1.8 \times 10^{4} \times 10^{24})$ = $5.7 \times 10^{-5}$ x $1.8 \times 10^{28}$ / $(4.2 \times 10^{7})$ = $2.4 \times 10^{7}$ cal/cm2.ns
For 1 keV: 300 cal/g ; for 1.5 keV: 340 cal/g [AD-A102 516 (1981)]

Planck's constant = h = $6.6261 \times 10^{-27}$ erg-s; E = hf
$\lambda$f = c, where c = the speed of light = $2.998 \times 10^{8}$ m/s, $\lambda$ = wavelength and f = frequency

Mach 1 is the speed of sound = 738 mph = 1,082 ft/s = 330 m/s (at STP)

1 calorie = 4.184 Joules = $2.4 \times 10^{6}$ ergs; 1 cal/cm$^2$ = $4.184 \times 10^{-2}$ MJ/m$^2$; 1 BTU = $2.5 \times 10^{2}$ calories
Energy/heat/work: 1 J = 1 N.m = 1 kg.m$^2$/s$^2$; 1 watt = 1 J/s (rate of power)
Pressure: 1 Pascal = 1 kg/m.s$^2$ ;$1.013 \times 10^{5}$ Pa = 1 bar = 14.7 psi = 1 atmosphere; 1 Gpa = $10^{9}$ Pa = 10 kbar; $10^{-6}$ dynes/cm2 = 1 bar
1,000 bar = 1 kbar; 1,000 kbar = 1 Mbar; 1,000 Mbar = 1 Gbar; 1,000 Gbar = 1 Tbar
tap = 0.1 N.s/m2; ktap = 100 N.s/m2
kip = 1,000 lbf = 4,448 N; kip/inch$^2$ (ksi) = $6.89 \times 10^{3}$ kPa

1 mil = 0.001"; 1 micron = $1 \times 10^{-6}$ meter

To double the volume of a sphere, multiply the radius or diameter by 1.266
To half the volume of a sphere, multiply the radius or diameter by 0.79

$10^{17}$ fissions Oralloy = $e^{39}$ = 39 generations = 15 MW for 0.2 seconds

Air Density (Nevada): 1.098 g/liter; Pressure at Earth's center = ~4 Mbar; Pressure at Sun's center: ~105 Mbar; temperature at Sun's core: 15.5 million °C.

1 barn = $1 \times 10^{-24}$ cm$^2$ = $1 \times 10^{-28}$ m$^2$

Newton's 2nd Law: momentum = mv = mass x velocity;
Force = ma = $\Delta$p/$\Delta$t = m $\Delta$v/$\Delta$t; a = $\Delta$v/$\Delta$t
Pressure = F/Area

$1 in 1945 = $17.87 in 2012; $1 in 1948 = $14.08 in 2012; $1 in 1996 = $1.45 in 2012


## Fission, Fusion, and Radiation

Average fission neutron energy = 2.0 MeV; Mean fission neutron energy = 1.6 MeV

velocity (km/s) = 1.38 x $10^4$ (v)½ , with v in MeV

3.5 x 1.6 MeV x 4.2 J/cal x $10^{-28}$ cm2 x 12 x.(134 x $10^3$) kt/kg x 2.62 x $10^{25}$ = 3.1 MeV/g T at Mbars of pressure & times of 10 shakes, and temperatures of tens of millions °C.

Pure fission devices produces $10^{23}$ neutrons/kt. [AD-A955 389 (1972)] [AD-A115 512 (1982)]
~100% fission H-Bomb produces 7.2 x $10^{23}$ 2 MeV neutrons/kt.
~100% fusion ("Clean") H-bombs produce up to 2 x $10^{24}$ neutrons/kt, about 7 times as many neutrons (and 14 MeV instead of an average of 0.5 MeV) as pure fission devices. [Cowan (1959)] [LAMS-2391 (1960)]
H-bombs produce 1.2 – 2.1 x $10^{24}$ neutrons/kt [AFSWP-1100 (1957)]
The 1952 Ivy Mike device produced Fermium-255 (17 neutrons added) out of the U238 present. [LAMS-2391 (1960)] ["Phys. Rev." 119(6): 2000-2004 (1960)]

1 Becquerel = 1 disintegration/s; 1 Curie = 3.7x$10^{10}$ disintegrations/s

Dosage 1 meter from a 2 x $10^{16}$ fission burst from Godiva (Oy) is approx. 60 rads from gamma rays, plus 740 rads from fast neutrons. [UCRL-ID-126119 (1959)]
1 ton yield of a fission bomb produces 40 rem at 400 yard distance. 1 rem = 1 rad/s
The mean free path of A-bomb gamma radiation is 400 yds. For the 22 kt 1951 Operation Ranger Fox Event, a gamma dose of 3,000 Roentgens was measured at 1,000 yds. The gamma dose of 1.2 kt Ranger Able was $10^9$ Roentgens from the origin; it was 4 x $10^8$ Roentgens at 400 yds. [LA-1228 (1951)]

Gamma dose in rem = $R_\gamma$ = (3 x $10^6$) (W) ($e^{-D/\lambda}$)/ $D^2$ ; where W = nuclear yield in tons; D = distance from the point of detonation in yards; λ = gamma mean free path = 350 yds. The prompt gamma dose for a "typical" (50/50 fusion/fission) TN weapon outside the atmosphere = 9 x $10^4$/$R^2$ rads (Si) per kt; where R is in km.

Neutron dose in rem = $R_n$ = (2.5 x l$0^7$) (W) ($e^{-D/\lambda}$)/ $D^2$ ; where W = nuclear yield in tons; D = distance from the point of detonation in yards; λ = neutron mean free path in yards = 250 yds. Density of air (nitrogen) = 4.02 x $10^{19}$ atoms/cm$^3$; Density of air (oxygen) = 1.07 x $10^{19}$ atoms/cm$^3$. Neutron flux for a low-yield pure fission weapon: $10^{23}$ neutrons/kt. Neutron flux for a "typical" (actually, for an unclassified 50/50 fusion/fission) TN weapon = (1.6 x $10^{12}$)/$R^2$ neutrons/cm$^2$ per kt; where R is in km. The neutron fluence with energy greater than 4 MeV is 0.167 of the total fluence.

Neutron/gamma dose in the atmosphere = (S/4π $r^2$) ($e^{-\mu m}$) , where S = neutron/gamma strength, r is the range, μ is the attenuation constant in cm$^2$/g (for 1 MeV gamma rays = 2.8 x $10^{-2}$ cm$^2$/g), and m is the air mass in g/cm$^2$ (1.22 mg/cm$^3$ at sea level, with an exponential decrease with increasing altitude). 10% of the neutrons produced are 10-15 MeV in a thermonuclear explosion (50/50 fusion/fission). [AD-A131 287 (1981)]

For a yield of 1 ton, the $LD_{50}$ (within 30 days) for total radiation (neutrons and gammas) is 150 yds. at least. For radiation acute illness effects will extend to ~225 yds. The emergency exposure level (100 rem or Roentgens equivalent) is 300 yds. For a 1 ton yield, the radiation dosage is 1,800 rem at 100 yds. [LA-2079-del (1956)]

1 Gray = Absorbed Radiation Dose = 1 Gy = 1 J/kg = $10^4$ ergs/g = 100 rads
1 rad = 2 x $10^9$ MeV/cm$^2$ in the case of prompt gamma rays and with element with atomic numbers less than 20. [AD-A955 403, p. 5-21 (1972)]; $LD_{50}$ = 650 rad plus or minus 150 rad. [LA-4350-MS (1969)]
100 rem = 1 Sievert per second = 1 J/kg.s
1 Roentgen = 1 R = absorbed Gamma dose = 6.77 x $10^7$ keV/cc air = 0.084 J/kg = 84 ergs/g = 2.58 x $10^{-4}$ Coulomb/kg (C/kg)
rep = Roentgen equivalent-physical; 1 rep = 83 erg/g tissue = 5.17 x $10^7$ MeV/g tissue

$R = (n)(6.023 \times 10^{23})(\sigma)(\rho)/A$ ; where R = number of neutron reactions (reactions/cm$^3$); n = total neutron flux/number of neutrons/cm$^2$; $\sigma$ = cross section for the neutron reaction (barns); $\rho$ = element's density (g/cm$^3$); A = atomic weight (g/mole). The neutron reaction mean free path (cm) = n/R

## Thermonuclear Fusion Reactions

Yield:

| | |
|---|---|
| Fusion of Deuterium (D + D): | 82.2 kt/kg |
| Fusion of Tritium + Deuterium: | 80.4 kt/kg |
| | 134 kt/kg Tritium |
| Fusion of Li6D (Li6 -> T, then D + T): | 64.6 kt/kg |
| Fusion of Li7D: | Uncertain:  ideally 6/7 of Li6D |

[Glasstone (1972)]

Li6 (n,α) T ; the triton generated has an energy above 2.7 MeV; if the triton undergoes fusion with a deuterium atom (D-T reaction) (which it does in an H-bomb), the neutron generated will have energy between 10.8 – 19.7 MeV. [Noshkin (2001)] (I.e., an average energy of 15.25 MeV.)

1 amu = $1.6597 \times 10^{-27}$ kg; neutron rest mass = 1.008982 amu = $1.6746 \times 10^{-27}$ kg

$v = c\,(1-(m_0)^2/m^2)^{.5}$
$\qquad m = 3.566 \times 10^{-30} + 1.6746 \times 10^{-27} = 1.6781657 \times 10^{-27}$
$\qquad$ for 2 MeV:  $c \times (1 - (1.6746 \times 10^{-27})^2/(1.6781657 \times 10^{-27})^2)^{.5}$

speed of light = c = $2.99 \times 10^{10}$ cm/s = 0.30 cm/ns
Fission neutron [2 MeV] speed:  0.065c = $1.95 \times 10^9$ cm/s
Fission neutron [0.5 MeV] speed:  0.033c = $9.8 \times 10^8$ cm/s
14 MeV neutron speed:  $\qquad$ 0.17c = $5.15 \times 10^9$ cm/s

Transfer of energy between collisions of particles of mass M1 and M2.  Partition of energy
$\qquad$ between two nuclear particles produced in a reaction is inverse to their relative mass.
Elastic scattering of a 14 MeV fusion neutron with:
$\qquad$ a deuteron acquires on average 9.4 MeV, but can acquire up to 12.5 MeV;
$\qquad$ a triton acquires on average 3.5 MeV, but can acquire up to 10.6 MeV;
$\qquad$ an He3 acquires on average 3.5 MeV, but can acquire up to 10.6 MeV;
$\qquad$ a proton acquires on average 7 MeV, but can acquire up to 10.6 MeV.

## Fissile Material Properties

Critical Mass of Spheres

| | Bare | Density (g/cm3) | 2" thick U | 1" thick Be |
|---|---|---|---|---|
| delta-Pu239/4.8 wt% Pu240 | 16.6 kg | 15.8 | | |
| alpha-Pu239/4.8 wt% Pu240 | 10.5 kg | 19.86 | | |
| 94% U235 | 49   kg | 18.8 | | |
| 93.5% U235 (Oralloy) | 52   kg | 18.8 | | |
| 93.2% U235 | 52.5 kg | 18.8 | | |
| U233 | 14.5 kg | 18.5 | | |

Total Fission of Oralloy (93.5% U235):  17.4 kt/kg
Total Fission of natural U (U238):  $\qquad$ 17.7 kt/kg
Total Fission of Weapons Grade Pu (1968 Production, 93% Pu):  17.7 kt/kg

Fast Fission (0.5 MeV) Cross Section ($\sigma$):     U235: 1.3 b ; 2.56 neutrons/fission
                                                  Pu239: 1.9 b; 3.08 neutrons/fission
                                                 U233: 2.1 b; 2.60 neutrons/fission

14 MeV Fission Cross Section:  U235: 2.18 b        4.1 n/fission
                                     U238  1.07 b        4.5 n/fission
                                     Pu239  2.55 b       4.9 n/fission

1972 Weapons-Grade delta-Pu: (94.5% Pu239 + 4.5% Pu240) + 1 wt.% gallium; $\rho$ = 15.8 g/cm3
                     alpha-Pu: 94.5% Pu239 + 4.5% Pu240; $\rho$ = 19.7 - 19.86 g/cm3

Weapons-Grade Plutonium: 94% Pu239 + 5.8% Pu240 + 0.2% Pu241 [ANRCP-1999-21 (1999)]
Watt/kg WG Pu: 2.63   [UCRL-ID-114164 (1993)]

Super-grade Plutonium: 2-3% Pu240, FY81
Super-grade Oralloy: 97% U235

Spontaneous Fission Rate (fission/s-kg):

U233 = < 0.19
U235 = 0.16           neutrons/spont. fission = 1.9
U238 = 5.5             n/SF = 2.0
Pu239 = 10.1         n/SF = 2.2
Pu240 = $4.5 \times 10^5$     n/SF = 2.2
Th232 = 0.041

|  | Half-life (y) | Activity (dpm) |
|---|---|---|
| Am241 | 433 | $7.4 \times 10^7$ |
| Np237 | $2.14 \times 10^6$ | $8.8 \times 10^6$ |
| Cm242 | .446 | $3.2 \times 10^9$ |
| Cm244 | 18.1 | $1.4 \times 10^7$ |

Spontaneous Neutron Generation:  Pu240 = 1,000,000 neutrons/s-kg
Weapon-grade Pu (4.5% Pu240) Neutron Generation:  45 n/ms-kg = 0.045 n/µs-kg
Atomic Batteries:  Heat Generation Pu238 = 567 W/kg; critical mass Pu238 = 9 kg

|  | Spont. Fission Neutron Rate (n/kg-s) | Heat (W/kg) |
|---|---|---|
| U233 | 1.3 | 0.27 |
| U235 | 0.41 |  |
| Pu239 | 3 | 1.9 |
| Pu240 | $1.0 \times 10^4$ | 7.0 |
| Pu238 | $2.6 \times 10^6$ | 567 |
| Am241 | 0.623 | 106 |
| Cm242 |  | 3000 |

[Lovins (1980)]

Mean free path for any neutron interaction [elastic, inelastic, fission, etc.] (cm):

|  | Density (g/cm3) | 1 MeV | 10 MeV |
|---|---|---|---|
| U235 | 18.8 | 3.0 | 3.4 |

| U238 | 18.8 | 2.3 | 3.0 |
| Pu239 | 19.7 | 3.0 | 3.0 |
| Lead | 11.4 | 6.4 | 5.0 |
| Tungsten | 19.3 | 2.3 | 3.2 |
| Beryllium | 1.84 | 4.5 | 10 |

Mean free path of 14 MeV neutrons for fission of U238, $\lambda$ = 7.6"

Tritium: 9,650 Curies/gram

Nuclear Fallout (50/50 fission/fusion device):

one hour after surface detonation, 550 MegaCuries/kt of fission yield, weighing 0.125 lbs or 2 oz.  [AD-A020 193, p. 3 (1975)] [AD-A955 403, p. 5-66 (1972)]
1 kt after 1 minute has produced $3 \times 10^4$ Megacuries.  After 1 day, fallout decay is 15 Megacuries/kt  [AD-A140 111, p. 28 (1983)]

1 day:  0.01 Curies less than zero time [Hicks (1984)]
50 days:  $4 \times 10^{-4}$ less Curies than zero time (long-lived isotopes) [Hicks (1984)]

"Best" fast neutron absorber:  Boron-10; Best neutron reflector:  BeO (a ceramic).

$B_{10}$ (n,$\alpha$) $Li_7$ , with the alpha particle having an energy of 2.8 MeV, and the Li7 atom having an energy of 2.3 MeV; (with a cross section of 3 barns at neutron energy of 0.5 MeV; and a very low cross section at 14 MeV)

$Li_7$ (n,$\alpha$) $He_4$; both alphas have an energy of 8.7 MeV.

Mean free path of fast fission neutron in air is about 250 yds.
Mean free path of fission gamma rays (3 x 2 MeV gamma) in air is about 350 yds.
1900 REM dose at 100 yds. from a 1 ton pure fission explosion  [LA-2079, p.47, 49 (1956)]
200 rads at 100 yds from a 1 ton (50/50) thermonuclear explosion.  [AD-A047 389 (1976)]

Neutron Bombs (JENDL 4.0):  neutron elastic scattering (total pretty much) for nitrogen:
 3 b      3 b      2 b      1 b
0.5 MeV 1 MeV  2 MeV   14 MeV

Beryllium elastic cross section about 3.3 b at 1 MeV, but only 1.5 b at 14 MeV.

Pu239:  mp. 641 ºC.; b.p. 3,232 ºC.;
Specific Heat of Pu;  32 cal/kg deg.C @ 25 ºC; 41 cal/kg ºC. @2000 ºK ;
Heat of fusion 2.82 kcal/kg; Heat of vaporization:  8.0 kcal/kg

*********
$I = I_o \times e^{-ux}$

Lambert's Law of Absorption; a measure of radiation attenuation.  $I_o$ is the original intensity in keV; I is the intensity, in keV, after passing through a thickness (in cm) x of a material whose mass attenuation factor is $\mu$.  Transmission factor is $I/I_o$ .

$\sigma$ = x-ray cross section (cm2/g) = $\mu/\rho$ ; $\mu$ = mass attenuation factor
$\rho$ = density (g/cm3)
$\sigma$ = 975 cm2/g Hg ; $\mu$ = 13.55 x 975 = 13,211/cm
$\rho$ = 11.85 g/cm3 thallium;  13.55 g/cm3 Hg

********
$I = 5$ keV x $e^{-13,211x}$

$4.5/5 = e^{-13,211x}$; $1 = e^{-x}$; $-8 \times 10^{-6} = -x$

$x = 8 \times 10^{-6}$ cm3 Hg $= 1.1 \times 10^{-4}$ g Hg

$(1.1 \times 10^{-4})/291$ moles; $1.1 \times 10^{-3} \times 22.4 = 24.6$ ml;

*******

(or for neutrons, $I = I_o \times e^{-\delta N x}$ ; where $\delta$ is the cross section in barns ($10^{-24}$ cm2); N number of nuclei/cm3; x = depth under the surface in cm)

$I = Io \times e^{-\mu/\rho}$, where I is the intensity at depth x (in cm) in keV
$= (4.5 \text{ keV}) / e^{975}$
$e^{20} = 6,400,000,000$

for u238/14 MeV = 1.07 b;  4 b for elastic, inelastic, (n,2n), (n,3n).
for u235/14 MeV = 2.2 b;
for Pu/14 MeV = 2.55 b; 3.3 b for elastic, inelastic, (n,2n), (n,3n).

for u238:
$N = N_o/e^{-(1.07 \times 10-24)(18.8 \times 6 \times 1023)(y)}$ cm
$\quad = N_o/e^{-2 \times 10-4 \times y}$ cm


## Thermal Conductivity

Thermal Conductivity of a material:   $(dQ/dt) = -\lambda A (dT/dx)$, where $(dQ/dt)$ is the time rate of heat conduction through any cross-sectional area A, of a thin slab of thickness dx, across which there exists a temperature difference dT, and $\lambda$ is the material's thermal conductivity.

| | | |
|---|---|---|
| BeO:  good thermal conductivity: | 330 W/ºC.m = 0.79 cal/s.ºC.cm | |
| BeO at 300 ºF: | 0.38 cal/s.ºC.cm | [CRC (1972)] |
| Be:    good thermal cond. | 200 W/ºC.m | |
| Pyrolytic Carbon:  good thermal cond. | 240 W/ºC.m = 0.57 cal/s.ºC.cm | |

| | | |
|---|---|---|
| Graphite: | 100 W/ºC.m = 0.24 cal/s.ºC.cm | |
| Pyrolytic graphite: | 0.93 | |
| Teflon: | 0.25 | [AD666 245, p. III-32 (1967)] |
| Hg | 8.3 | |


## Rocket Force Equation

Specific Impulse = $I_{sp}$ = F/(dm/dt) = thrust (lbs)/propellant flow rate (lbs/s).  The units of the Specific Impulse is therefore in seconds.

$F = I_{sp} \times (dm/dt)$ ; where F = force (kg); $I_{sp}$ = specific impulse of propellant (seconds); (dm/dt) = mass of propellant gas discharge per unit time (kg/s) = P x S x r; where P = propellant density (kg/cm3), S = surface area of propellant being burned (cm2), and r = burn rate of propellant (cm/s).

Therefore, $F = I_{sp} \times P \times S \times r$

*********

The change in burnout velocity with respect to missile weight may be approximated by the equation:

$d(V_{BO})/d(W_{TOT}) = g (I_{sp}) (1/W_{BO} - 1/W_{TOT})$

where $V_{BO}$ = velocity at burnout; $W_{BO}$ = weight at burnout; $W_{TOT}$ = weight at launch; g = gravitational constant  [AD530 958 (1974)]


## Heat of Combustion

$CO + 1/2\ O_2 \rightarrow CO_2$;   94.3 kcal/28g
$C + O_2 \rightarrow CO_2$; 171 kcal/12 g
$H_2 + \frac{1}{2}\ O_2 \rightarrow H_2O$;    57.8 kcal/2g


## Densities

Hydrogen: b.p. -252.87 ºC; liquid density = 0.089 g/cm3
Deuterium: b.p. -249.49 ºC; liquid density = 0.169 g/cm3
Tritium:  b,p. -248.12 ºC;  liquid density = 0.26 g/cm3
He3:  liquid density = 0.095 g/cm3; atomic number density = $1.9 \times 10^{22}$ $cm^{-3}$
LiD, ρ = 0.88 g/cm3; LiH, ρ = 0.77 g/cm3; LiT, ρ = 0.989 g/cm3
Be, ρ = 1.84 g/cm3; m.p. 2,340 F; BeO (beryllium oxide), ρ = 2.69 g/cm3; BeH2 (beryllium hydride), ρ = 0.65 g/cm3; 0.78 g/cm3 also reported
C (graphite) ρ = 2.22 g/cm3 ; m.p. 3,700 ºC.
Al (aluminum), ρ = 2.84 g/cm3
WC (tungsten carbide), ρ = 14.7 g/cm3; m.p. 2,630 ºC.; ~92% W alloy, ρ = 17.4 g/cm3
UH3 (uranium hydride) ρ = 10.95 g/cm3
PuH3 (plutonium hydride) ρ = 10.4 g/cm3


## Refractories

W (Tungsten):  m.p. 3410 ºC
C (graphite):  m.p. 3700 ºC; b.p. 4827 ºC.; Specific Heat .17 cal/(g)(ºC)
TaC (tantalum carbide):  mp. 3880/4730 ºC.


## Velocity of Sound

Air:        ~330 m/s
Beryllium:  12,890 m/s
Tungsten:   5410 m/s


## High Explosives

|  | Crystal density (g/cm3) | Det. Velocity (km/s) | C-J Pressure (kbar) |
|---|---|---|---|
| TNT | 1.65 | 7.0 | 190 |
| RDX | 1.81 | 8.3 | 360 |
| PETN | 1.78 | 8.4 | 320 |
| HMX | 1.91 | 9.1 | 420 |
| Comp. B (60/40) | 1.73 | 8.0 | 263 |
| Cyclotol 75/25 | 1.74 | 8.2 | 312 |
| Octol (HMX/TNT) 76/24 | 1.82 | 8.4 | 345 |
| PBX 9404 | 1.85 | 8.8 | 360 |
| PBX 9502 | 1.90 | 7.7 | 285 |
| TATB | 1.85 | 7.9 | 291 |

| Cast Baratol (24% TNT, 76% barium nitrate) | 2.62 | 4.9 | 160 |
|---|---|---|---|
| Pressed Baratol | | 5.9 | |
| Boracitol (40% TNT, 60% boric acid) | 1.59 | 4.7 | 87 |

[WASH-1037REV (1972)]  [AFWL-TR-66-12 (1966)]  [Glasstone (1972)]  [Crawford (1976)]
[LA-UR-81-1067 (1981)]  [LA-9053-MS (1981)]  [LA-UR-81-1067 (1981)]  [LA-13014-H (1995)]  [AD383 490 (1967)]

## Plastic-Bonded Explosives (PBX)

C-J Pressure = $(2.5 \times 10^{-6}) \rho D^2$ kbar   ["ETI Handbook", p. 43]

| | |
|---|---|
| PBX 9010 | 90 wt. % RDX, 10% Kel-F 3700 elastomer, 1.78 g/cm3; C-J pressure 320 kbar |
| PBX 9011 | 90 wt. % HMX, 10% Estane elastomer; 1.795 g/cm3; C-J pressure ~298 kbar |
| PBX 9404 | 94% HMX, 3% nitrocellulose, and 3% CEF (phosphate ester) plasticizer, and 0.1% diphenylamine stabilizer; 1.866 g/cm3; vacuum pressed to high density from powder; C-J pressure 367 kbar |
| PBX 9407 | detonator booster pellet |
| PBX 9501 | 95% HMX + 2.5% Estane 5703-F1 (a polyester urethane elastomer), and 2.5% DNPAF nitroplasticizer as binders + 0.1% diphenylamine stabilizer; 1.855 g/cm3; 8.8 km/s; vacuum pressed at 300 tons at 85 ºC.; developed as a replacement for PBX 9404 which had stability problems; C-J pressure ~375 kbar |
| PBX 9502 | IHE; 95% TATB, 5% Kel-F 800 binder; 1.89 g/cm3; C-J press. 285 kbar; $V_{det}$ = 7.7 km/s |
| LX-04 | 85 wt % HMX, 15% Viton A; first used in 1962 as replacement HE for the Sergeant SRBM [Drell & Perurifoy (1994)] [DASA-1220 (1962)] |
| LX-10-1 | 94.5% HMX, 5.5% Viton A; 1.87 g/cm3; 8.84 km/s; pressed PBX |
| LX-11 | 80% wt. % HMX, 20% fluoroelastomer; 1.884 g/cm3;  C-J Pressure 330 kbar |
| LX-14 | 95.5 wt. % HMX, 4.5% Estane; 1.85 g/cm3; 8.8 km/s; pressed PBX |
| LX-17-1 | IHE; 92.5% TATB + 7.5% Kel-F-800 polymer binder; 1.9 g/cm3.  C-J Press. 275 kbar; $V_{det}$ = 7.6 km/s; used in B83 aerial bomb, W84 GLCM, and W87 Peacemaker ICBM MIRV. [UCRL-JC-133766 (1999)] |

## An Essential Delivery Systems Bibliography

Cochran, Thomas et al., "Nuclear Weapon Databook, Volume 1: U.S. Nuclear Forces and Capabilities". Cambridge, Massacheusetts: Ballinger Publishing Co. (1984). [Though dated, this volume is still **the** standard reference on U.S. nuclear weapons.]

DNA 4501F. "Physics of High-Altitude Nuclear Burst Effects", Defense Nuclear Agency, AD-A068 541 (December 1977).

Francis, Sybil. "Warhead Politics: Livermore and the Competitive System of Nuclear Weapon Design", UCRL-LR-124754, Livermore, CA: LLNL (June 1995) [her MIT Ph.D. thesis].

Gibson, James N. "Nuclear Weapons of the United States". Atglen, Pennsylvania: Schiffer Publishing (1996).

Hansen, Chuck. "U.S. Nuclear Weapons: The Secret History". New York: Orion (1988). [superseded by Hansen (1994)]

Hansen, Chuck. "The Swords of Armageddon: U.S. Nuclear Weapons Development Since 1945". (CD or microfiche). Sunnyvale, California: Chuckelea Publications (1994).

RDD-7. "Restricted Data Declassification Decisions 1946 to the Present", 7th edition. Germantown, Maryland: U.S. Department Of Energy, Office of Declassification (January 1, 2001).

## General Bibliography

Addington, Larry H. "The Patterns of War Since the Eighteenth Century". Bloomington, Indiana: Indiana University Press (1984).

Aldridge, Robert. "Nuclear Empire". Vancouver, Canada: New Star Books (1989).

Anderson, Kevin, "Explosives", in Carleone, Joseph, ed. "Tactical Missile Warheads". Washington, D.C.: AIAA (1993).

Anderson, Martin. "Reagan's Secret War". NY: Crow Publishers (2009).

Arkin, William M. and Fieldhouse, Richard W. "Nuclear Battlefields: Global Links in the Arms Race". Cambridge, Massachusetts: Ballinger Publishing Co. (1985).

Baker, David. "The Rocket". NY: Crown Publishers (1978).

Bethe, Hans A. "The Road from Los Alamos". NY: American Institute of Physics (1991).

Bille, Matt and Lishock, Erika. "The First Space Race: Launching the World's First Satellites". College Station, Texas: Texas A&M University Press (2004).

Bonds, Ray, ed. "The Illustrated Directory of Modern Weapons". London: Salamander Books (1986).

Boyne, Walter J. "Beyond the Horizons: The Lockheed Story". NY: St. Martin's Press (1998).

Brauch, ed. "Military Technology, Armament Dynamics, and Disarmament" (1989).

Braun, Werner von and Ordway (1975).

Bruce, James B. "CIA Studies in Intelligence", "The Consequence of Permissive Neglect". (>= 2002)

Brugioni, Dino A. "Eyes in the Sky: Eisenhower, the CIA, and Cold War Aerial Espionage". Annapolis, Maryland: Naval Institute Press (2010).

Burrows, William E. "By Any Means Necessary: America's Secret Air War in the Cold War". NY: Farrar, Strauss and Giroux (2001).

Carleone, Joseph, ed. "Tactical Missile Warheads", Vol 155. Washington, D.C.: AIAA (1993).

Chaisson, Eric J. "The Hubble Wars". NY: HarperCollins Publisher, Inc. (1994).

Chayes, Abram and Wiesner, Jerome B., eds. "ABM: An Evaluation of the Decision to Deploy an Antiballistic Missile System". NY: Harper & Row (1969).

Chrzanowski, Edward J. "Active Radar Electronic Countermeasures". Norwood, Mass.: Artech House (1990).

Cirincione, Joseph. "Bomb Scare: The History & Future of Nuclear Weapons". NY: Columbia University Press (2008).

Clark, John D. "Ignition! An Informal History of Liquid Rocket Propellants". New Brunswick, New Jersey: Rutgers University Press (1972).

Clearwater, John. " 'Just Dummies': Cruise Missile Testing in Canada," Calgary, Alberta, Canada: University of Calgary Press (2006).

Coates, James and Kilian, Michael. "Heavy Losses: The Dangerous Decline of American Defense". NY: Viking Penguin Inc. (1985).

Dalgleish, D. Douglas and Schweikart, Larry. "Trident". Carbondale, Illinois: Southern Illinois University Press (1984).

De Maeseneer, Guido. "Peenemünde: The Extraordinary Story of Hitler's Secret Weapons V-1 and V-2". Vancouver, Canada: AJ Publishing Co. (2001).

De Maeseneer, Guido. "The First 100 Years: Tsiolkovsky to Sputnik, 1857-1957". Vancouver, Canada: AJ Publishing Co. (2001).

Dickson, Paul. "Sputnik: The Launch of the Space Race". Toronto: Macfarlane Walter & Ross (2001).

Dorman, Bernie, et al. "Aerojet: The Creative Company". San Dimas, California: Aerojet History Group (1995).

Doyle, Stephen H., ed. "History of Liquid Rocket Engine Development in the United States 1955- 1980". San Diego, California: AAS History Series, Vol. 13, Proceedings of an American Astronautical Society History Colloquium (1992).

Drell, Sidney D. "Facing the Threat of Nuclear Weapons". Seattle, Washington: University of Washington Press (1983).

Dunnigan, James F. and Nofi, Albert A. "Dirty Little Secrets: Military Information You're Not Supposed to Know". NY: William Morrow & Co. (1990).

Eggleston, Wilfrid. "Canada's Nuclear Story". Toronto: Clarke, Irwin & Co. (1965).

Emme, Eugene. Perry, Robert L. and Miles, Wyndham D, in "The History of Rocket Technology". Detroit, Michigan: Wayne State University (1964).

Ford, Brian J. "Secret Weapons: Technology, Science & the Race to Win World War II". Oxford, U.K.: Osprey Publishing (2011).

Foster, Caxton C. and Iberall, Thea. "Computer Architecture", 3rd edition. NY: Van Nostrand Reinhold Co. (1985).

Gardiner, Robert, ed. "Navies in the Nuclear Age: Warships since 1945". Annapolis, Maryland: Naval Institute Press (1993)

Gertz, Bill. "Betrayal". Washington, D.C.: Regnery Publishing Inc. (1999).

Graham Jr., Thomas and Hansen, Keith A. "Spy Satellites and Other Intelligence Technologies that Changed History". Seattle, Washington: University of Washington Press (2007).

Greenwood, John T., ed. "Milestones of Aviation". NY: MacMillan Publishing Co.(1989).

Greenwood, Ted. "Making the MIRV: A Study of Defense Decision Making". Cambridge, Massachusetts: Ballinger Publishing Co. (1975).

Gunston, Bill. "Jet Bombers: From the Messerschmitt ME 263 to the Stealth B-2". London: Osprey Aerospace (1993). ** [Lengthy and detailed descriptions of B-47/B-52 et al.]

Hadley, Arthur T. "The Straw Giant". NY: Avon Books (1971, 1984, 1986, 1987).

Hambling, David. "Weapons Grade: How Modern Warfare Gave Birth to Our High-Tech World". NY: Carroll & Graf Publishers (2005).

Harclerode, Peter. "Warfare". London: Macmillan Publishers Ltd. (2000).

Harwood, William B. "Raise Heaven and Earth: The Story of Martin Marietta People and Their Pioneering Achievements". NY: Simon & Schuster (1993).

Heppenheimer, T.A. "Flight: A History of Aviation in Photographs". Richmond Hill, Ontario, Canada: Firefly Books Ltd. (2004).

Hoag, D.G. "Ballistic-Missile Guidance" in Feld, B.T. et al. "Impact of New Technologies on the Arms Race". A Pugwash Monograph. Cambridge, Mass.: MIT Press (1971).

Hodge, Nathan and Weinberger, Sharon. "A Nuclear Family Vacation: Travels in the World of Atomic Weaponry". NY: Bloomsbury (2008).

Hogg, Ian V. "German Secret Weapons of the Second World War". Mechanicsburg, Pennsylvania: Stackpole Books (1999).

Howard-White, F.B. "Nickel: An Historical Review". Toronto, Canada: Longmans Canada Ltd. (1963).

Hunley, J.D. "The Development of Propulsion Technology for U.S. Space-Launch Vehicles, 1926-1991". College Station, Tx: Texas A&M University Press (2007).

Ippolito, Jr., Thomas Dominic. "Effects of Variation of Uranium Enrichment on Nuclear Submarine Reactor Design". MIT, M.Sc. Thesis (1990).

Ivanov, B.N. "Fundamentals of Physics". Moscow: Mir Publishers (1989).

Johnson, Brian. "The Secret War". NY: Methuen (1978).

Kaplan, Fred. "The Wizards of Armageddon". NY: Simon & Schuster (1983).

Karp, Aaron. "Ballistic Missile Proliferation". SIPRI (Stockholm Internation Peace Research Institute). NY: Oxford University Press (1996).

Kaufmann III, William J. "Universe", 3rd edition. NY: W.H. Freeman and Company (1985).

Keeney, L. Douglas. "15 Minutes: General Curtis LeMay and the Countdown to Nuclear Annihilation". NY: St. Martin's Press (2011). [doesn't cover nuclear weapons from a design perspective, but is an excellent analytical history of SAC and its history]

Kennedy Willliam V. Col. "The Intelligence War". London: Salamander Books Ltd. (1983).

Loh, W.H.T. "Re-entry and Planetary Entry Physics and Technology II: Advanced Concepts, Experiments, Guidance-Control and Technology" NY: Springer-Verlag (1968).

Lohm W.H.T. "Re-entry and Planetary Entry Physics and Technology", Vol. 3. NY: Springer-Verlag (1968).

Long, Eric F. et al. "At the Controls: The Smithsonian National Air and Space Museum Book of Cockpits". Erin, Ontario, Canada: Boston Mills Press (2001).

Lord, M.G. "Astro Turf: The Private Life of Rocket Science". NY: Walker & Co. (2005).

Mackenzie, Donald. "Inventing Accuracy: A Historical Sociology of Nuclear Missile Guidance". Cambridge, Mass.: MIT Press (1990).

McKay, Paul. "Atomic Accomplice: How Canada deals in deadly deceit". (2009).

McNamara, Robert S. "Blundering into Disaster". NY: Pantheon (1986).

McNamara, Robert S. "In Retrospect: The Tragedy and Lessons of Vietnam". NY: Vintage (1995).

Miller, Gerald E. "Stockpile: The Story Behind 10,000 Strategic Nuclear Weapons." Annapolis, Maryland: Naval Institute Press (2010).

Morgan, Christopher et al. "Draper at 25", Cambridge, Mass.: The Charles Stark Draper Laboratory (1998).

Muller, Richard A. "Physics for Future Presidents". NY: W.W. Norton & Co. (2008).

Jr. Nero, Anthony V. "A Guidebook to Nuclear Reactors". Los Angeles: University of California Press (1979).

Nijboer, Donald and Patterson, Dan. "Cockpits of the Cold War". Erin, Ontario, Canada: Boston Mills Press (2003).

Oberdorfer, Don. "The Turn: From the Cold War to the New Era". NY: Simon & Schuster (1992).

Page, Leigh and Adams Jr., Norman Ilsley. "Electrodynamics". NY: D. Van Nostrand Co, Inc. (1940).

Paglen, Trevor. "Blank Spots on the Map". NY: Penguin Group (2010).

Panofsky, W.K.H. "Arms Control and SALT II". Seattle, Washington: University of Washington Press (1979).

Peebles, Curtis. "The Corona Project: America's First Spy Satellites". Annapolis, Maryland: Naval Institute Press (1997).

Poirier, Bernard. "Witness to the End: Cold War Revelations: 1959-1969". NY: University Press of America (2000).

Polmar, Norman and Noot, Jurrien. "Submarines of the Russian and Soviet Navies, 1718-1990". Annapolis, Maryland: Naval Institute Press (1991).

Powaski, Ronald A. "Return to Armageddon: The United States and the Nuclear Arms Race, 1981-1999". NY: Oxford University Press (2000).

Priest, Dana and Arkin, William M. "Top Secret America: The Rise of the New American Security State". NY: Little, Brown and Company (2011).

Raytheon Missile Systems. "Tomahawk Cruise Missile" Fact Sheet (2002).

Raviv, Dan and Melman, Yossi. "Spies Against Armageddon: Inside Israel's Secret Wars". Sea Cliff, NY: Levant Books (2012).

Reed. Thomas C. "At the Abyss: An Insider's History of the Cold War". NY: Random House (2004).

Richelson, Jeffrey T. "America's Space Sentinels: DSP Satellites and National Security". Lawrence, Kansas: University Press of Kansas (1999).

Richelson, Jeffrey T. "The Wizards of Langley". Boulder, Colorado: Westview Press (2001).

Richelson, Jeffrey T. "Defusing Armageddon: Inside NEST, America's Secret Nuclear Bomb Squad". NY: W.W. Norton & Co., Inc. (2009).

Rosenbaum, Ron. "How the End Begins: The Road to a Nuclear World War III". NY: Simon & Schuster (2011).

Scarry, Elaine. "Rule of Law, Misrule of Men". Cambridge, Massachusetts: Boston Review Book (2010).

Scheer, Robert. "Thinking Tuna Fish, Talking Death: Essays on the Pornography of Power". NY: Farrar, Straus, and Giroux (1988).

Scott, Bill. "Inside the Stealth Bomber: The B-2 Story". Blue Ridge Summit, Pennsylvania: TAB Aero Books (1991).

Schwartz, Stephen I. "Atomic Audit". Washington, D.C.: Brookings Institute Press (1998).

Sheehan, Neil. "A Fiery Peace in a Cold War". NY: Random House (2009).

Smith, F.G. Walton. "The Seas in Motion". NY: Thomas Y. Crowell Co. (1973).

Spangenburg, Ray, and Moser, Diane Kit. "Modern Science 1896-1945". NY: Facts on File Science Library (2004).

Spinardi, Graham. "From Polaris to Trident: the Development of U.S. Fleet Ballistic Missile Technology". NY: Cambridge University Press (1994).

Stine, G. Harry. "ICBM: The Making of the Weapon that Changed the World". NY: Orion Books (1991).

Sutton, George P. "Rocket Propulsion Elements: An Introduction to the Engineering of Rockets", 6th edition. NY: John Wiley & Sons, Inc. (1992).

Talbott, Strobe. "Deadly Gambits: The Reagan Administration and the Stalemate in Nuclear Arms Control". NY: Alfred A. Knopf (1984).

Taylor, J.H. and Taylor, W.R. "Missiles of the World". London: Ian Allan Ltd. (1972).

Thompson, Kenneth W., ed. "Kosta Tsipis on the Arms Race: A Collection of Critical Essays". NY: University Press of America (1987).

Van Cleave, William. "Nuclear Technology and Weapons" in Lawrence, Robert and Laurus, Joel, eds. "Nuclear Proliferation Phase II". NY: University Press of Kansas (1974).

Van Creveld, Martin. "The Age of Airpower". NY: Public Affairs (2011).

Wells, Jr., Samuel J. "Nuclear Weapons and European Security during the Cold War", in Hogan, Michael J., ed. "The End of the Cold War". London: Cambridge University Press (1992).

West, Nigel. "The SIGINT Secrets". NY: William Morrow and Co., Inc. (1988).

Wiencek, Henry. "The Lords of Japan". Chicago: Stonehenge Press (1982).

Yenne, Bill. "U.S. Guided Missiles: The Definitive Reference Guide." Manchester, U.K.: Crecy Publishing Ltd. (2012).

Zaloga, Steven J. "Target America: The Soviet Union and the Nuclear Arms Race, 1945-1964" (1993).

Zaloga, Steven J. "The Kremlin's Nuclear Sword: The Rise and Fall of Russia's Strategic Nuclear Forces, 1945-2000". Washington, D.C.: Smithsonian Institution Press (2002).

Zimmerman, Robert. "The Universe in a Mirror: The Saga of the Hubble Space Telescope and the Visionaries Who Built It". Princeton, New Jersey: Princeton University Press (2008).

## Papers and Periodicals

AIAA 92-2759. Burnett, Jimmy and McCain, J. Wayne. "Fast Burn Booster Technology", American Institute of Aeronautics and Astronautics (1992).

Badenhorst, Nick P. "The Bomb, the Missile, and the Future", in "Armed Forces" [South Africa], September 1993, p. 25-32; October 1993, p. 26-33; November 1993, p. 27-34; December-January, p. 22-26.

Barlow, Jeffrey G. "Moscow and the Peace Offensive". (1982).

Bethe, Hans, Garwin, Richard L. et al. "Space-based Ballistic-Missile Defense". "Sci. Am." 251(4):39-49 (October 1984).

"Bell System Technical Journal", Special Supplement, "Safeguard Data-Processing System" (1975).

Bergen, Delmar. Interview by G. Spinadi. Los Alamos, N. Mex. (December 18, 1990).

Blair, Bruce. "Keeping Presidents in the Nuclear Dark". "Bruce Blair's Nuclear Column Home" on the Internet (February 11, 2004).

Brand, Stewart. "Founding Father". Paul Baran interview in "Wired" Magazine, March 2001, p. 144-153.

Bunn, Matthew. "Technology of Ballistic Missile Reentry Vehicles", Report No. 11. Cambridge, Massachusetts: MIT, Program in Science and Technology in International Security (March 1984).

Burr, William, ed. "The Creation of SIOP-62", National Security Archive Electronic Briefing Book No. 130 (posted on web July 13, 2004). (gwu.edu/~nsarchiv)

Burton, Mark. "MX Missile to Peacekeeper: The R&D Process Fulfiled", "Peak of Flight Newsletter", Issue 333, April 22, 2013, e-zine on Internet Web (2013).

"Congressional Record", June 9, 2005 (House), Page H4340-H4345.

Day, Dwayne A. "Early Reentry Vehicles: Blunt Bodies and Ablatives", www.centennialofflight.gov (2003).

Drell, Sidney D. "Physics and U.S. National Security", "Rev. Mod. Phys." 71(2):S460-S470 (1999).

Esser, Alexandra S. "Fractured German-American Relations: The Effects of Presidential Personality". Dept. of Government, Franklin & Marshall College (2006).

Garwin, Richard L. "Why China Won't Build U.S. Warheads", "Arms Control Today" (April/May 1999).

Kamal, Seyed Arif and Mirza, Ashab. "The Multi-Stage-Q System and the Inverse-Q System for Possible Application in Satellite-Launch Vehicle". Fourth International Bhurban Conference on Applied Sciences and Technologies, Bhurban, Pakistan (2005).

Leitenberg, Milton. "Studies on Military R&D and Weapon Development", in "Case Study 3: The Origin of MIRV". Swedish Ministry of Defense (1984).

--. "Swords into Plowshares". "Life" magazine, p. 26-32 (November, 1987).

Lodal, Jan M. "Assuring Strategic Stability", in "Foreign Affairs" 54(3):3 (1976).

Morgan, Christopher et al. "Draper at 25". Cambridge, Mass.: Charles Stark Draper Laboratory (1998).

Norris, Robert S. et al. "United States Secretly Deployed Nuclear Bombs in 27 Countries and Territories During Cold War". NRDC, National Security Archive Electronic Briefing Book No. 20 (on George Washington University gwu.edu/NSA web site) (October 20, 1999).

Rosenbaum, Ron. "The Subterranean World of the Bomb". "Harper's" magazine (March 1978).

Richelson, Jeffrey. "The Keyhole Satellite Program", "J. Strat. Stud.", (June 1984).

Richelson, Jeffrey T. "Defusing Armageddon: Inside NEST, America's Secret Nuclear Bomb Squad". NY: W.W. Norton & Co., Inc. (2009).

Romig, Mary F. "The Physics of Meteor Entry". P-2902. Santa Monica, Ca.: The RAND Corp. (1964).

Smith, Levering et al. "Innovative Engineering in the Trident Missile Development", in "The Bridge" [Nat. Acad. of Engineering] 10(2):10-19 (1980).

wikipedia.org, "Q-Guidance" (2012).

**Declassified and Other Government Documents**

AD042 398. "Titan II IGS Category II Maintenance, Logistics, Reliability and Readiness", AC Sparkplug Division, General Motors Corp., Ballistic Systems Division, Air Force Systems Command, Norton AFB, San Bernadino, California (1963).

AD046 714. "Project MX-776 Quarterly Progress Report," BMPR-38, Bell Aircraft Corp. (1954).

AD113 976. "Project MX-776 Quarterly Progrees Report", Report No. 56-981-021-46, Bell Aircraft Corp. (Sept. 30, 1956).

AD158 516. Redman, E.J. and Pasiuk, L. "Pressure Distributions on an ABMA Jupiter Nose Cone...", Navord Report 4486, U.S. Naval Ordnance Laboratory, White Oak, Maryland (1957).

AD264 660. Robinson, R.B. and Uzdarwin, R.J. "Investigation of Stress-Corrosion Cracking of High-Strength Alloys". Aerojet-General for Frankford Arsenal (1961).

AD268 311. Jaffee, R.I. et al. "Fabrication of Tungsten for Solid-Propellant Rocket Nozzles," DMIC Memorandum 136, Defense Metals Information Center, Battelle Memorial Institute, Columbus, Ohio (1961).

AD268 330. "Development of Deep Drawn – One Piece High Performance Rocket Motor Case," General Report No. 11, Lyon Inc., Detroit, Michigan, Frankford Arsenal (1961).

AD269 349. MacPherson, B.M. and Beaver, W.W. "Fusion Welding of Beryllium," WADD Technical Report 60-917, Brush Beryllium Co., Aeronautical Systems Division, Air Force System Command, Wright-Patterson Air Force Base, Ohio (1961).

AD295 888. Zagorites, H.A. And Sinclair, K.F. "A Continuous Scan Scintillation System for the Inspection of Tungsten Billets". USNRDL-TR-602. San Francisco, Ca.: U.S. Naval Radiological Defense Laboratory (1962).

AD313 420. Russell, Jr., William J. et al. "Effects of High Altitude Nuclear Detonations on High Frequency Communications." AFSWP-1104, HQ Defense Atomic Support Agency (1959).

AD315 467. Liimatainen, T.M., ed. "Progress in Miniaturization and Microminiaturization", Report No. PR-60-1, Diamond Ordnance Fuze Laboratories, Department of the Army (Dec. 1959).

AD331 804. "Report of Beryllium Committee", National Academy of Sciences, National Research Committee (1962).

AD338 603. "Characteristics of Tactical, Strategic and Research Missiles." Convair, San Diego (1957).

AD339 910. "Evaluation of the A3D-1 Aircraft for Special Weapons Delivery Capability." WT-1334, Operation Redwing – Project 5.8, Bureau of Aeronautics, Department of the Navy (1956).

AD353 247. Johannesson, Karl R. "Least Dispersion of the Mark 12 Re-Entry Vehicle", Technical Report 178, Air Weather Service, U.S. Air Force (1964).

AD354 894. Bryan, John H. "Chaff Countermeasures and Air Defense Radar Design" Technical Report 6, Stanford Research Institute, Redstone Arsenal, Alabama (April 1959) [originally classified Secret].

AD355 925. "Final Report. Satellite Interception System Feasibility Report" Vol. 2, ARPA (1963).

AD368 640. "High Chamber Pressure Rocketry Program", Report AFRPL-TR-65-191, Aerojet-General Co., Rocket Propulsion Laboratory, Research and Technology Division, Air Force Systems Command, U.S. Air Force, Edwards, California (December 1965).

AD379 893. Skaggs, G.A. et al. "Reentry Phenomena Observed Using High-Frequency Radar", NRL Report 6507, Naval Research Laboratory, Washington, D.C. (Feb. 1967).

AD383 490. Higuera, R.L. and Menz, F.L. "Warhead Studies for the Period Ending 30 June 1967", IED-B7, Naval Ordnance Department Corona, Corona, California (July 1967).

AD388 603. Hanson, C.M. "Characteristics of Tactical, Strategic, and Research Missiles", ZM-486, Convair, General Dynamics, San Diego (April 1957).

AD392 561. Olcott, E.L. "Development of Pyrolytic Graphite Coatings for Rocket Nozzles", AFRPL-TR-68-145, Atlantic Research Co. (1968).

AD392 715. Jones, Roy E. et al. "Demonstration of Advanced Post-Boost Propulsion Subsystems Final Report", Technical Report AFRPL-TR-68-126, Aerojet-General Corp., Air Force Rocket Propulstion Laboratory, Air Force Systems Command, Edwards AFB (1968).

AD401 115. Batchelor, James D. "Improvement of the Usefulness of Pyrolytic Graphite in Rocket Motor Applications", Atlantic Research Corp., Alexandria, Virginia, U.S. Army, Ordnance Materials Research Office, Watertown Arsenal, Watertown, Massachusetts (1963).

AD405 855. Morris, Deane N and Benson, P. "Data for ICBM Re-entry Trajectories." RM-3475-ARPA, Rand Corp., Santa Monica, California, Defense Advanced Research Agency (1963).

AD421 893. Chang, P.K. "Analysis of the Aerodynamic Ablation of a Metal Sphere", U.S. Naval Research Laboratory (1963).

AD464 318. McCreight, L.R. et al. "A Survery of the State of the Art of Ceramic and Graphite Fibers". General Electric, AFML-TR-65-105 (1965).

AD465 896. Burleson, W.G. and Reynalds, R.A. "Theoretical Effects of Reentry Aerodynamic Heating on the External Skin Structure of AMRAD, Experiment No. 1," Report No. RS-TR-65-3, Advanced Research Projects Agency, U.S. Army Missile Command, Redstone Arsenal, Alabama (April 1965).

AD477 394. Lindberg, H.E. et al. "Hardening Technology Studies: Response of Reentry Vehicle-Type Shells to Blast Loads," LMSC-B130200, Vol. IV-C, Lockheed Missile & Space Co., Ballistic Systems Division, Norton Air Force Base, California (1965).

AD483 114. Axtell, Robert C. and Potter, Richard M. "Some Contributions of the Vela Satellite Program in Space Research". SSD-TR-65-142, ARPA (DoD), Space Systems Division, Air Force Systems Command, Los Angeles Air Force Station, Los Angeles, California (December1965).

AD530 958. Moorhead, Seth B. "Refractory Air Vane and Refractory Material Research and Development, Task II – Refractory Materials for a Thrust Vector Control Valve," AMMRC CTR 74-47 Martin Marietta Aerospace, Orlando, Florida, Army Materials and Mechanics Research Center, Watertown, Massachusetts (1974).

AD600 907. Morgan, H. "Leading Edge and Nose Cap Materials – Pyrolytic Graphite," SVE Memo No. 125-142, U.S. Air Force BSD-ARDC (1961).

AD664 961. Bukalov, V.M., et al. "Atomic-Powered Submarine Design" (1967); translated from the Russian.

AD688 074. Katz, R. Nathan and Gazzara, Charles P. "Recent Developments in Pyrolytic Graphite", AMMRC MS 69-01, Ceramics Research Laboratory, Army Materials and Mechanics Research Center, Watertown, Massachusetts (1969).

AD701 545. Williams, F.A. et al. "Fundamental Aspects of Solid Propellant Motors", AGARDograph 116, The Advisory Group for Aerospace Research and Development, NATO (October 1969).

AD731 662. Glover, L.S. and Hagan, J.C. "The Motion of Ballistic Missiles", Technical Memorandum TG 1164, The John Hopkins University, Applied Physics Laboratory, Strategic Systems Project Office, Department of the Navy (July 1971).

AD742 765. Larsen, J.V. and Smith T.G. "Carbon Fiber Structure". NOLTR 71-165, Silver Spring, MD: Naval Ordnance Laboratory (1971).

AD746 008. Allen, Douglas. "Laboratory Conversion and State Descriptor of the D-17B Computer." Air Force Institute of Technology, Wright-Patterson Air Force Base, Ohio (1972).

AD760 757. Cicirelli, Raymond. "Reutilization of the Minuteman Guidance Computer as a Numerial/ Process Controller (1973).

AD763 495. Seyrfai, Kahlil. "Engineering Design Handbook on Infrared Military Systems". Part 1, U.S. Army Materiel Command (1971).

AD767 570. "Active Nosetip Evaluation Study," SAMSO-TR-73-74, 20050203232, McDonnell Douglas Astronautics Co., Air Force Systems Command, Space and Missile Systems Organization, Los Angeles, California (1972).

AD773 168. Diefendorf, Russell J. et al. "The Relationships of Structure to Properties in Graphite Fibers, Part II", Rensselaer Polytechnic Institute, AF Materials Laboratory (1973).

AD777 244. Reguli, Dennis C. "Conversion of the D37C Computer for General Purpose Applications", Air Force Institute of Technology (1974).

AD783 200. Baetz, Jay G. "Experimental Evaluation of Coated Graphite ICBM Rocket Nozzle Throat Inserts," SAMSO-TR-74-158, Aerospace Corp., Space and Missile Systems Organization (1974).

AD783 849. Payne, William F. "Pyrolytic Graphite Coated Throat Inserts", AFRPL-TR-74-42, Air Force Rocket Propulsion Laboratory, Edwards AFB, California (1974).

AD786 472. Thompson, Grant and Day, Evan E. "Quarterly Progress Report No. 9, April – June 1974, Development of HTPB Propellant for Ballistic Missiles", AFRPL-TR-74-44, Thiokol Corp., Air Force Rocket Propulsion Laboratory, Edwards AFB (July 1974).

AD787 040. "Aerospace Structural Adhesives", NMAB-300, National Materials Advisory Board (NAS-NAE), ODDR&E Department of Defense (July 1974).

AD818 416. "Review of Alloys and Fabricating Methods Used for Tactical Missile Motor Cases." DMIC Memorandum 224, Defense Metals Information Center, Battelle Memorial Institute, Columbus, Ohio (1967).

AD824 697. Clark, Thomas J. "Development of Manufacturing Methods for Producing Pyrolytic Graphite in Various Shapes and Structures", Air Force Materials Laboratory, Wright-Patterson AFB, Ohio (1967).

AD830 267. "Engineering Design Handbook: Ballistic Missile Series Structures" (1963).

AD850 572.  Schmidt Jr., A.E. "The Dynamic Response of Graphitic Materials" (1969).

AD851 310.  Sorenson, H.W. "Range and Guidance Accuracy Capability of the Altas Missile System," Report No. ZN-7-366 (June 1960).

AD857 651.  Branigan, John E. "Long-Term Storability of Propellant Tankage and Components", Technical Report AFRPL-TR-69-82, Air Force Rocket Propulsion Laboratory, Air Force Systems Command, Edwards AFB, California (1969).

AD858 125.  Hayes, Arthur F. "Production of Forged Beryllium Conical Structural Shapes," Technical Report AFML-TR-69-168, Ladish Co., Air Force Materials Laboratory, Air Force Systems Command, Wright-Patterson Air Force Base, Ohio (1969).

AD-A002 696.  Hung, J.C. et al. "IMU Self-Alignment Techniques." Battelle Columbus Laboratories, Army Missile Research, Development and Engineering Laboratory, Redstone Arsenal, Alabama (1974).

AD-A014 428.  Stanbery, Charles E. "SRAM Explosive Components Surveillance Program Summary Report and FY74 Service Life Estimate," ASD-TR-75-4, Volume 1 (1975).

AD-A017 242.  "Missile Manufacturing Technology Conference..." Army Materiel Command, Alexandria, Virginia (1975).

AD-A032 600.  Hlavinka, Duane K. "Lessons Learned:  Production Restart of a Major Weapons System," Study Project Report PMC 76-1, Fort Belvoir, Virginia (1976).

AD-A033 540.  Stetson, J.R. and Schutzler, J.C. "Evaluation of Carbon-Carbon Composite Nosetip Materials," AMMRC CTR 76-34, Prototype Development Associates, Inc., Army Materials and Mechanics Research Center, Watertown, Massachusetts (1976).

AD-A033 726.  Ossin, Archie and Kendall, Paul. "Evaluation of CONAP Concept for Advanced ABM Nose-tips". AAMRC CTR 76-38, U.S. Army Materiel Command, Alexandria, Virginia (November 1976).

AD-A036 610.  "A Special Report on Oceanographic Program Planning for the Deputy..." Office of the Oceanographer of the Navy, Alexandria, Virginia (1964).

AD-A036 741.  Wright, Alfred C. "USAF Propellant Handbooks, Nitric Acid/Nitrogen Tetroxide Oxidizers, Volume II", AFRPL-TR-76-76, Martin Marietta Corp., Denver, Colorado, Air Force Rocket Propulsion Laboratory, Director of Science and Technology, Air Force Systems Command, Edwards AFB, California (February 1977).

AD-A043 064.  "Evaluation of the Virgin Mechanical and Thermal Properties of AVCO 3DCC Heatshield Materials." Southern Research Institute, DNA 4212F, Defense Nuclear Agency (1977).

AD-A047 389.  Gritzner, M.L. et al. "Radiation Environment from Tactical Nuclear Weapons", DNA 4267F, Science Applications Inc., Huntsville, Alabama, Defense Nuclear Agency, Washington, D.C. (July 1976).

AD-A055 778.  "Strap-Down Inertial Systems". AGARD-LS-95 (1978).

AD-A057 762.  Fedoroff and Sheffield. "Encyclopedia of Explosives and Related Items", Vol. 8.  PATR 2700. NJ:  Picatinny Arsenal (1978).

AD-A058 574.  Records, Jr. Louis Russell.  "Control Aspects of Highly Constrained Guidance Techniques" (1979).

AD-A063 901.  Hunter, Joe S.  "Lance Configuration Q-Flex Accelerometer Design Verification", Technical Report T-78-91, U.S. Army Missile Research and Development Command, Redstone Arsenal, Alabama (September 1978).

AD-A065 881.  Mlinar, Anthony J.  "Application of Learning Curves of Aircraft Produced at More Than One Location to the F-16 Leightweight Fighter."  AFIT/GSM/SM/78S-16 (1978).

AD-A066 217.  Smith, David H.  "Flame Nosetip Recovery Vehicle Flight Test Series."  DNA 4654F, Prototype Development Associates, Inc., Santa Ana, California, Defense Nuclear Agency (1978).

AD-A066 814.  Wong, Anthony K. et al.  "Fracture Toughness and Stress Corrosion Characteristics of Ultrahigh-Strength 4340 Steel – Summary Review" (1979).

AD-A067 516.  White, H.V. and Hung, J.C.  "Gyrocompassing Error Analysis for Pershing II Inertial Measurement Unit", Redstone Arsenal (1978).

AD-A076 485.  Garcia, W. et al.  "Composite Material Application to the Mk 12A RV Midbay Substructure." AMMRC TR 79-51, General Dynamics Corp., Convair Division, Army Materials and Mechanics Research Center (1979).

AD-A080 547.  Hove, John E. and Gowen, Leo F.  "Retrospective Study of Selected Dod Materials and Structures Research and Development Programs Phase I:  Case History Data Collection", IDA Paper P-1391, Institute for Defense Analyses (March 1979).

AD-A090 151.  Langley, Maj. R. Warren and Billings, Dana.  "Multiple, Independently Targeted Reentry Vehicle (MIRV) Targeting Models", SRL-TR-80-0017, Frank J. Seiler Research Laboratory, USAF Academy, Colorado, U.S. Air Force, Air Force Systems Command (August 1980).

AD-A090 649. Whitman, Robert B.  "Astroinertial Navigation for Cruise Applications".  Hawthorne, California:  Northrop Corporation (1980).

AD-A090 650.  Gates, Robert V. and Hall, Mark G.  "SLBM Fire Control Computational Algorithms In Support of Stellar Inertial Guidance", Dahlgren, VA:  Naval Surface Weapons Center (1980).

AD-A098 593.  "General Test Report, Production Lot Sampling:  Minuteman III/Mark 12A Reentry Vehicle Carbon-Carbon Nosetip Production", AVCO Systems Division, Document AVSD-0125-80-CR, Dept. of the Air Force, Ballistic Missile Office, Norton AFB, San Bernadino, California (14 April 1980).

AD-A098 595.  "MM III / Mk 12A Reentry Vehicle Carbon/Carbon Nosetip Production Program,"   AVCO Document AVSD-0114-80-CR, AVCO Corp., Dept. of the Air Force, Ballistic Missile Office (1980).

AD-A102 516.  Triebes, K. & Liu, G.  "Titanium Response to a Nuclear Radiation Environment", DNA 5596F, Acurex Corp. Aerotherm Division, Mountain View, California, Defense Nuclear Agency, Washington, D.C. (1981).

AD-A108 511.  White, H.V. et al.  "Software Features Applicable to Inertial Measurement Unit Self-Alignment."  Technical Report RG-81-11, U.S. Army Missile Command, Redstone Arsenal, Alabama (1980).

AD-A111 686.  MacNaughton, Michael G.  "Toxicology of High Energy Fuels", AFAMRL-TR-81-136, 20060630472, Air Force Aerospace Medical Research Laboratory, Wright-Patterson AFB, Ohio (Dec. 1981).

AD-A112 526. "The Shock and Vibration Bulletin", Bulletin 43, Part 4, The Shock and Vibration Center, Naval Research Laboratory, Washington, D.C., Office of the Director of Defense Research and Engineering (June 1973).

AD-A113 928. "Low Cost Heatshield Materials". Alabama: Redstone Arsenal (1982).

AD-A115 691. Clitchclow, Carl L. and William, Ronald G. "A Simulation Model for Analyzing Reentry Vehicle/AntiBallistic Missile Engagements" (1982).

AD-A116 828. Schmidt, D.L. and Craig, R.D. "Advanced Carbon Fabric/Phenolics for Thermal Protection Applications", AFWAL-TR-82-4136, Materials Laboratory, AF Wright Aeronautical Laboratories, AF Systems Command, Wright-Patterson AFB, Ohio (1982).

AD-A120 491. Hopkins, J.C. "The Development of Strategic Air Command". Office of the Historian, Headquarters Strategic Air Command (1982).

AD-A121 264. Hunter, Joe S. et al. "Lance Q-Flex Accelerometer Qualification Test Program", Techncial Report RG-82-6, U.S. Army Missile Command, Redstone Arsenal, Alabama (March 1982).

AD-A127 320. "Lance Handbook, Firing Team Leader's" U.S. Army Field Artillery School, Weapons Department, Fort Sill, Oklahoma (March 1983).

AD-A139 157. "Justification of Estimates For Fiscal Year 1985 and 1986", Department of the Navy, Weapons Procurement, Navy (1984).

AD-A141 082. Conroy, E.H. et al. "The Joint Cruise Missiles Project: An Acquisition History – Appendices." Rand Corp., Santa Monica, California, The Joint Cruise Missiles Project Office (1982).

AD-A142 764. "Geodesy for the Layman." DMA-TR-80-003, Defense Mapping Agency (1983).

AD-A148 828. "Manufacturing Methods & Technology, Program Plan, CY 1984." U.S. Army Materiel Command (1984).

AD-A149 597. Haas, W.R. and Prince, S. "Atmospheric Dispersal of Hypergolic Liquid Rocket Fuels," Volume 1, ESL-TR-84-18, Martin Marietta Aerospace, Denver, Colorado, Engineering and Services Laboratory, Air Force Engineering and Servcies Center, Tyndall AFB, Florida (1984).

AD-A158 180. Barattino, William John. "Coupled Radiation Transport/Thermal Analysis of the Radiation Shield for a Space Nuclear Reactor," AFIT/CI/NR 85-530, dissertation (1985).

AD-A162 646. Werrell, Kenneth P. "The Evolution of the Cruise Missile", Air University Press, Alabama: Maxwell Air Force Base (1985).

AD-A174 324. Dunbar, W.G. and Silverman, S.W. "Cruise Missile Power System," AFWAL-TR-85-2113, Boeing, Seattle, Washington (1986).

AD-A180 927. Siebein, Kerry N. "Mirostructural Characterization of a Cobalt-Free Maraging Steel, Vasco Max T-250". MTL TR 87-12, U.S. Army Materials Technology Laboratory (1987).

AD-A190 476. "Measures and Trends U.S. and U.S.S.R. Strategic Force Effectiveness". DNA 4640-F, Santa Fe Corporation, Alexandria, Virginia, Defense Nuclear Agency, Washington, D.C. (July 1978).

AD-A197 073. "SAC Needs a Few Good Men and Woman". Ebbs, Raymond E., Maxwell Air Force Base, Alabama (1988).

AD-A197 162.  Cousins, Frank W.  "The Anatomy of the Gyroscope, Part I", NATO AGARD Monograph No. 313 (1988).

AD-A199 356.  "Design Method in Solid Rocket Motors," AGARD-LS-150 (Revised), Advisory Group for Aerospace Research and Development, NATO, Neuilly sur Seine, France (1988).

AD-A206 251.  Nauta, Frans.  "Logistics Implications of Maneuver Warfare."  Report IR702R4, Logistics Management Institute (1988).

AD-A209 273.  Knacc, Marcelle Size.  "Encyclopedia of U.S. Air Force Aircraft and Missile System, Volume 2, Post-World War II Bombers 1945-1973", Office of Air Force History, U.S. Air Force, Washington, D.C. (1988).

AD-A210 747.  Atkins, Jr., Issac and Dibben, Mark.  "Evaluation of Motor Exhaust and Liner Combustion By-Products".  AFOEHL Report No. 80-038EH0087EAC, AF Occupational and Environmental Health Laboratory, Brooks AFB, Texas (May 1989).

AD-A214 665.  Watt, K. et al.  "TLAM/N in the Nuclear Reserve Force".  DNA-TR-84-84, Science Applications International Corporation (1983).

AD-A221 595.  "The Anatomy of the Gyroscope, Part 3", NATO AGARD Monograph No. 313.

AD-A224 584.  Bauer, Ernest.  "Physics of High-Temperature Air, Part 1, Basics", IDA Document D-487, Alexandria, Virginia:  Institute for Defense Analyses (1990).

AD-A233 379.  Checkel, Jeffrey.  "An Analysis of Soviet Military Writings on U.S. Reentry Vehicle Technology, 1965-1983".  Soviet Security Studies Working Group, MIT (1986).

AD-A228 013.  Kinas, Ernest N. and Hickey Jr., Charles F.  "Mechanical Property Characterization of Thick Wall Ti-6Al-6V-2Sn Forging".  Watertown, Mass:  U.S. Army Materials Technology Laboratory (1990).

AD-A229 778.  Bauer, Ernest.  "Physics of High-Temperature Air.  Part II.  Applications".  IDA Doc. D-626, Institute for Defense Analyses (August 1990).

AD-A231 552.  Finke, Reinald G.  "Calculation of Reentry Vehicle Temperature History", IDA Paper P-2395, Institute for Defense Analysis (1990).

AD-A235 080.  Day, Lt. Col. Lowell L.  "Eye Spy:  The Utility of Strategic Satellite Reconnaisance", MSc thesis,  School of Advanced Military Studies, U.S. Army Command and General Staff College, Fort Leavenworth, Kansas (1990).

AD-A238 593.  "Risk Assessment for Emergency Planning Related to Nuclear Weapons Accidents", SAIC-85-1849, Defense Nuclear Agency (1985).

AD-A241 725.  Van Atta, Richard H. et al.  "DARPA Technical Accomplishments".  Volume 2, Institute for Defense Analysis (April 1991).

AD-A250 424.  Lyon, Michael J.  "Introduction to Rocket Propulsion," Technical Report RD-PR-91-17, U.S. Army Missile Command, Redstone Arsenal, Alabama (December 1991).

AD-A254 602.  Mitchell, Eddie.  "Apogee, Perigree, and Recovery", RAND Note N3103-A (1991).

AD-A257 717.  Gannon, M.W.  "Cruise Missile Proliferation," PhD thesis, Naval PostGraduate School, Monterey, California (1992).

AD-A258 264.  Joyce, Brian D. and Proppert Patrick E.  "Life-Cycle Cost of Alternative ICBM Second Stage Designs", M.Sc. Thesis, AFIT/GCA/LSG/92S-5 (September 1992).

AD-A272 447.  Bauer, Ernest.  "Standard Atmospheres for Engagement Modeling," IDA Document D-1197, Institute for Defense Analyses, Alexandria, Virginia, Ballistic Missile Defense Organization (1993).

AD-A279 703.  Clapp, William G.  "Space Fundamentals for the War Fighter", Newport, R.I.:  Naval War College (1994).

AD-A286 599.  "Mild Wind Series, Minute Steak Event, Project Officers Report", DNA (1972).

AD-A307 510.  Oliver, J.W.  "Burst Test of a First-Stage Polaris A-3 Filament-Wound Configuration-X Chamber", Report No. B-235, Aerojet-General Corp., Dept. of Defense, Plastics Technical Evaluation Center, Picatinny Arsenal, Dover, New Jersey (August 1963).

AD-A309 937.  Vick, Alan et al.  "Enhancing Air Power's Against Light Infantry Targets".  RAND/MR-697-AF (1996).

AD-A314 441.  Terry, R.E.  "Lithium Hydride Debris Shields for Plasma Radiation Sources," NRL/MR/6720—96-7868, 19960924 102, Naval Research Laboratory, Washington, D.C. (1996).

AD-A318 763.  Morrison, Dr. Alfred M. & Vamos, Dr. John S.  "The Reentry Systems Application Program (RSAP)."  Naval Surface Warfare Center, Dahlgren Division (1996).

AD-A325 314.  Schmidt, Donald L.  "Carbon-Carbon Composites (CCC) -- a Historical Perspective", WL-TR-96-4107, Materials Directorate, Wright Laboratory, Air Force Materiel Command, Wright-Patterson Air Force Base, Ohio (1996).

AD-A328 596.  Watson, Robert J.  "Into the Missile Age, 1956 – 1960," Vol IV, Historical Office, Office of the Secretary of Defense, Washington, D.C., 19970825 148 (1997).

AD-A353 633.  Schake, Kurt Wayne.  "Strategic Frontier:  American Bomber Bases Overseas, 1950 – 1960", 19980915 005, Department of History, Norwegian University of Science and Technology, U.S. Department of the Air Force, AFIT/CIA, WP AFB, Ohio (January1998).

AD-A363 899.  "Effectiveness of the Minuteman II Stage III Refurbishment Program," Air Force Center for Studies and Analyses, HQ USAF, Asst. Chief of Staff (1985).

AD-A386 542.  Hopkins, R. et al.  "The Silicon Oscillating Accelerometer:  A MEMS Inertial Instrument for Strategic Missile Guidance", 20010201 041, Charles Stark Draper Laboratory, Inc. (1998).

AD-A387 318.  Boezer, Gordon et al.  "Handbook of Energetic Materials for Weapons Systems Including Ballistic and Cruise Missiles", IDA Document D-1703, Institute for Defense Analyses, 20010309 022, Defense Technology Security Administration (September 1995).

AD-A398 731.  Warmbrod, John D.  "Calculated Results for the Transient Heating and Melting Process of Glass Shield…", NASA TN D-1643, 20020111 032, National Aeronautics and Space Administration, Washington, D.C., George C. Marshall Space Flight Center, Huntsville, Alabama (1963).

AD-A399 207.  McCullough, Roy.  "Missiles at the Cape."  ERDC-CERL SR-01-22, U.S. Army Corps of Engineers (2001).

AD-A406 104.  Umholtz, Philip D.  "The History of Solid Rocket Propulsion and Aerojet".

AD-A425 146. Kuentzmann, P. "Introduction to Solid Rocket Propulsion," RTO-EN-023, Office National d'Etudes et de Recherches Aerospatiales, NATO (2004).

AD-A434 326. Satterfield, P. and Akens, D. "Army Ordnance Satellite Program" (1958).

AD-A434 478. "Development of the Corporal: The Embryo of the Army Missile Program", Vol. 2, Historical Monograph No. 4, Army Missile Command, Huntsville, Alabama (1961).

AD-A439 957. Neufeld, Jacob. "The Development of Ballistic Missiles in the United States Air Force 1945-1960." Air Force Historical Studies Office, Washington, D.C. (1990).

AD-A440 094. Neufeld, Jacob et al. "Technology and the Air Force: A Retrospective Assessment," Air Force Historical Studies Office, U.S. Air Force, Washington, D.C. (1997).

AD-A440 289. Albrecht, Meredith M. "The Effect of Aerodynamic Surfaces Versus Thrust Maneuvers on Reentry Vehicles", M.Sc. Thesis, AFIT/GAE/ENY/05-S01, Department of the Air Force, Air University, Air Force Institute of Technology, Wright-Patterson AFB, Ohio. (2005).

AD-A528 970. Marchese, Joseph, ed. " 'Breakthrough' Technologies Developed by the Air Force Research Laboratory and Its Predecessors". Air Force Research Laboratory History Program (2005).

AD-A511 972. "AU-18 Space Primer", Alabama: Air University Press, Maxwell AFB (2009).

AD-A531 197. Conrad, Edward E. et al. "Collateral Damage to Satellites from an EMP Attack." DTRA-IR-1022, 20101029076 (2010).

AD-A532 565. Hodgson, Quentin E. "Deciding to Buy: Civil-Military Relations and Major Weapons Programs", Carlisle, Pennsylvania: U.S. Army War College, Strategic Studies Institute (2010).

AD-A586 733. Bragg, James W. "Development of the Corporal: the Embryo of the Army Missile Program", Volume 1, Historical Monograph No. 4, Army Ballistic Missile Agency, Redstone Arsenal, Alabama (April 1961).

AD-A602 158. Meilinger, Phillip S. "Bomber: The Formation and Early Years of the Strategic Air Command," Air University Press, Air Force Research Institute, Maxwell Air Force Base, Alabama (2012).

AD-A606 621. Perry, Robert. "A History of Satellite Reconaissance", Volume IIB, SAMOS E5 and E6 (1973).

AD-A952 021. Hall, Kimball P. and Bastress, Karl E. "Burning Rate Control Factors in Solid Propellants", Dept. of Aeronautical Engineering, Princeton University, Advanced Research Projects Agency (ARPA), Dept. of the Navy, Office of Naval Research (August 1, 1961).

AD-A955 389. Dolan, Philip J.,ed. "Capabilities of Nuclear Weapons, Part 2, Damage Criteria", "Chapter 5: Nuclear Radiation Phenomena". Defense Nuclear Agency (1 July 1972).

AD-A955 391. Dolan, Philip J., ed. "Capabilities of Nuclear Weapons, Part 2, Damage Criteria, "Chapter 7: EMP Phenomena". (1972).

AD-A955 400. Dolan, Philip J., ed. "Capabilities of Nuclear Weapons, Part, Damage Criteria, "Chapter 16: "Damage to Missiles". (1972).

AD-A955 400. Dolan, Philip J.,ed. "Capabilities of Nuclear Weapons, Part 2, Damage Criteria", "Damage to Missiles", Chapter 16, DNA (July 1, 1972).

AD-A955 403. Dolan, Philip J. "Capabilities of Nuclear Weapons", Part 1, "Phenomenology", DNA EM-1, SRI International (1972).

AD-A995 454. Elliott, G.P. et al. "Operation Hardtack, Project 9.3a, Operation of Missile Carrier for Very-High-Altitude Nuclear Detonations," WT-1657(EX), Defense Nuclear Agency (1959).

AD-A995 502. Murray, W.W. "Operation Dominic, Shot Sword Fish, Scientific Director's Summary Report," WT-2007(EX), Defense Nuclear Agency (1963).

AD-B007 377. Karpp, R.R. and Predebon, W.W. "Calculations of Fragment Velocities from Naturally Fragmenting Munitions", Memorandum Report No. 2509, U.S. Ballistic Research Laboratories, Aberdeen Proving Grounds, Maryland (July 1975).

AD-B060 927. „Lightweight Advanced Post-Boost Vehicle Propulsion Feed System Development", AFRPL-TR-81-26, Rocketdyne Division, Rockwell International, Air Force Rocket Propulsion Laboratory, Edwards AFB, California (1981).

AAMRC CTR 74-47. Moorhead, Seth B. "Refractory Air Vane and Refractory Material Research and Development Task II". Alexandria, Virginia: U.S. Army Materiel Command (July 1974).

ADC 020 813. "The Feasibility of Population Targeting", DNA, Science Applications Inc. (1979).

ADC 021 800. Wainstain, Leonard. "The Origins of the Cruise Missile: A Study in Technological Innovation". IDA Paper P-1386, Institute for Defense Analysis (1980).

Adkins, Lance K. "ICBM Command and Control". "High Frontier" Vol. 2, No.3, p.38-43, Air Force Space Command (2006).

Air Force Historical Research Center, "History of the Strategic Air Command, 1 January 1958 – 30 June 1958", Historical Study No. 73, Volume 1, Air Force Historical Research Center, Maxwell AFB, Alabama (1959).

AWE, Atomic Weapons Establishment, Nuclear Effects Group, "Electro-Magnetic Pulse (EMP)", (from web page awe.co.uk.; probably around 2005).

"A Study of SR-71 Utility for Post-Strike Reconnaissance" (1971).

Bair, Jeffrey A. "An Examination of I.C.B.M. Development within the United States from 1952 to 1965", thesis, U.S. Army Command and General Staff College (2003).

BC73-60, "The Lance Rocket Engine", Rocketdyne Division, Rockwell International (1973).

Be1018. ORF50569. Memorandum from Hightower, E.C. to Hibbs, R.F. "BOB, A-76 Study Ceramic Parts" (1969).

Be2536. ORF94536. ORO111769. West, R.M. "Memo to File, Requisition H-9130". (1961).

Bennett, Bruce W. "A RAND Note: Assessing the Capabilities of Strategic Nuclear Forces: The Limits of Current Methods". N-1441-NA (1980).

Buchonnet, Daniel. "MIRV: A Brief History of Minuteman and Multiple Reentry Vehicles", COVD- 1571, LLNL (February 1976).

Buck, R.A. "Re: Detonator Shock Test," Case No. 438.00, Ref. Sym 1612 (284), Project No. TM-223 D-2 (1955).

Bullard, John W. "History of the Redstone Missile System". Redstone Arsenal, Alabama (1965).

Burnett, J. ""Fast Burn Booster Technology". Thiokol/AIAA. AIAA 92-2759. U.S.A.D.C., Huntville, Alabama (1992).

Cagle, Mary T. "History of the Sergeant Weapon System". Redstone Arsenal, Alabama (1971).

Caywood, E.R. "History of the Joint Strategic Target Planning Staff", SAC Historical Staff (1967).

CIA. "Soviet Work on Radar Cross Section Reduction Applicable to a Future Stealth Program", CIA, Directorate of Intelligence (1984).

CIA. "The Balance of Nuclear Forces in Central Europe", SR 78-10004 (1978).

Congressional Budget Office. "The Trident II Missile Test Program: Implications for Arms Control", U.S. Congress, Congressional Budget Office (1987).

"Countermeasures", Appendix F, "The Reentry Heating of Submunitions" (ca. 2000).

Cousins, T. DTIC #AD-A212 748, "The Use of the Computer Code ATR to Relate DREO Experimental Results to Nuclear Battlefield Results (U)". Ottawa, Canada: Defense Research Establishment Ottawa (February 1989).

Cummings, C. "STRAT-X, Vol. 16 Reaction -- USSR Strategy", Institute for Defense Analyses Report R-1222. (1967).

DASA-1229. "Electromagnetic Blackout Guide" Vol. 1, DTIC No. AD 427 020 (1961).

DASA-1335. Richmond, Donald and White, Clayton. "A Tentative Estimation of Man's Tolerance to Overpressures From Air Blast". (1962)

DASAST, "Memo from Chairman, JCS to Asst. to the Sec. of Defense (Atomic Energy)". DASA (1962).

Day, Dwayne A. "Early Reentry Vehicles: Blunt Bodies and Ablatives". Centennialofflight.gov/essay/ Evolution_of_Technology/reentry/Tech17.htm off of web (September 2003).

DC 61-835. "Revision to Weapons Development Budget Submission for FY 1963," UCRL, Livermore (1961).

DeLucas, Kathy. "Blast from the Past: Lab Scientists Receive Vindication", "LASL, Daily News Bulletin" (Friday July 11, 1997).

"Demonstrated Destruction of Nuclear Weapons; Field Test FT-34; Final Report, Volume 1", Field Operations, Weapons Evalutation and Control Bureau, U.S. Arms Control and Disarmament Agency (Jan. 1969).

DNA 5350F-3. "Prelaunch Survivability of Ground Launched Cruise Missile", Vol. 3 (1980).

DNA-6147T. Jackson, Victor J. "Dynamic Retargeting of U.S. Strategic Forces", The RAND Corporation, DTIC No. AD-A222 191 (1982).

Drake, James. "Design -- Land Mobile System" Strat-X, vol. 4, Institute for Defense Analysis Report R-1222 (1967).

Elliott, J.P. et al. WT-1657(EX). Operation Hardtack, Project 9.3a. "Operation of Missile Carrier for Very-High-Altitude Nuclear Detonations", DTIC No. AD-A995 454 (1959).

"Environmental Impact Statement Peacekeeper Missile System Deactivation and Dismantlement", Air Force Center for Environmental Excellence, Brooks AFB, Texas (December 2000).

"FY 2013 Congressional Budget Justification", Volume 1, "National Intelligence Program Summary", Office of the Director of National Intelligence and the United States Intelligence Community, (Top Secret/Special Intelligence/Talent-Keyhole) (February 2012).

GACAEC, 35th, "Minutes of the 35th Meeting of the General Advisory Committee to the U.S. Atomic Energy Commission" (May 14-16,1953).

GACAEC, 41st, "Minutes of the 41st Meeting of the General Advisory Committee to the U.S. Atomic Energy Commission" (July 12-15, 1954).

GACAEC, 50th, "Minutes of the 50th Meeting of the General Advisory Committee to the U.S. Atomic Energy Commission" (July 16-18, 1956).

GACAEC, 75th, "Minutes of the 75th Meeting of the General Advisory Committee to the U.S. Atomic Energy Commission" (July 13-15, 1961).

Glover, L.S and Hagan, J.C. "The Motion of Ballistic Missiles". Johns Hopkins University, Applied Physics Laboratory, AD731 662 (1971).

Goldberg. "History of the Strategic Arms Competition 1945-1972, Part I". Office of the Secretary of Defense, Historical Office (1981).

Goldberg. "History of the Strategic Arms Competition 1945-1972, Part II" Office of the Secretary of Defense, Historical Office (1981).

Gray, D.C. to Taylor, F.J., handwritten note on a Memo, "XW-47 Program". (Dec. 11, 1958).

Grimwood and Strow, "History of the Jupiter Missile System", U.S. Army Ordnance Missile Command (27 July 1962).

"History of AFSWP, 1947-1954", Volume 5 – 1952, Chapter 3.

"History of the Custody and Deployment of Nuclear Weapons, July 1945 Through September 1977", Office of the Assistant to the Secretary of Defense (Atomic Energy). (1978).

"History of the Phase Out of Large Yield Weapons," Systems Analysis (September 14, 1967).

"History of Strategic Air Command January – June 1968", Vol. 1, Narrative, Historical Study No. 112, History and Research Division, HQ SAC (February 1969).

"History of Strategic Air Command FY 1969," Historical Study No. 116, Vol. 1, Office of the Historian, HQ SAC (1970).

HQ SAC "History of the Joint Strategic Target Planning Staff: Background and Preparation of SIOP-62", History and Research Division (undated).

"ICBM Security Classification Guide", Department of the Air Force, ICBM System Program Office, HQ Ogden Air Logistics Center, Hill Air Force Base, Utah (30 September 1997). [Obtained from cryptome.org in 2006.]

LA-1659. Lemons, J.F. and Lewis, W.B. "Predicted Structure and Density of Beryllium Hydride". (1953).

LA-2079-del. Harris et al. "Contamination Hazard From Accidental Non-Critical Detonation of Small Atomic Devices". (September 1956).

LA-11401. Pollock, Raymond. "A Short History of the U.S. Nuclear Stockpile: 1945-1985". LANL (January 2, 1991).

LA-13014-H. Dobratz, Brigitta. "The Insensitive High Explosive Triaminotrinitrobenzene (TATB): Development and Characterization – 1888 to 1994". (August 1995).

LA-3181-MS. Dickinson, James M. "High-Deformation Rate Extrusion of Thin-Walled Tungsten Hemispheres". (August 1, 1964).

LA-3385-MS. Bell, George I. "Neutron Blanket Calculations for Thermonuclear Reactors". (1965).

LA-3417-MS. Hoerlin, H. "Air Fluorescence Excited by High-Altitude Nuclear Explosions". Los Alamos, New Mexico: Los Alamos Scientific Laboratory (April 1965).

LA-4350-MS. "Proceedings of the Tactical Nuclear Weapons Symposium" (1969).

LAMS-2391. Cowan, George A. "Nuclear Explosions as Neutron Sources", LASL (May 1959).

LASL-79-84. Barasch, Guy E. "Light Flash Produced by an Atmospheric Nuclear Explosion" (November 1979).

Lemmer, George F. "The Air Force and Strategic Deterrence, 1951-1960". Air Force Historical Division Liaison Office (December 1967).

"Minuteman Weapon System History and Description", ICBM System Program Office, Hill AFB, Utah (July 2001).

Murray, J.P. "Cross-Reference Sheet, Subject: Beryllium Recycle and Lithium Tetraborate Development", NV0319796 (July 20, 1960).

Murray, Thomas, memo Lewis Strauss, "Increasing U.S. Weapons Testing Capability by Continuous Testing" (March 28, 1957).

Narducci, Dr. Henry M. "Strategic Air Command and the Alert Program: A Brief History," Office of the Historian, HQ Strategic Air Command, Offutt Air Force Base, Nebraska (1988).

National Counterintelligence Center et al. "Foreign Collection Against the Department of Energy". (1998).

National Park Service. "History of Minuteman Missile Sites"; downloaded off web at nps.gov/mimi /history/srs/history.htm (ca. 1995).

National Park Service. "The Missile Plains: Front-line of America's Cold War – Historical Resource Study" (2003); downloaded off web at nps.gov/mimi.

National Reconnaisance Office. "A History of Satellite Reconnaisance", Volume 1 – Corono (1973).

NAVWAG (Naval Warfare Analysis Group) Study No. 1. "Introduction of the Fleet Ballistic Missile into Service" (30 January 1957).

Neufeld, Jacob, "USAF Ballistic Missile Programs 1969-1970", Office of Air Force History (1990).

Nitze, Paul. U.S. Senate, Committee for Foreign Relations, "Hearing for the SALT II Treaty", Part I, p. 458 (1979).

NRL Report 5097, SER: 00808/RD. "A Satellite and Space Vehicle Program for the Next Steps Beyond the Present Vanguard Program", Washington, D.C.: Naval Research Laboratory, AD339 967 (December 10, 1957).

Office of the Secretary of Defense, Historical Office. "History of the Strategic Arms Competition, 1945-1972", Part I (March 1981) (was Top Secret).

OTA. "MX Missile Basing", Office of Technology Assessment (1981).

Power, Thomas S. "History of the Joint Strategic Target Planning Staff: Preparation of SIOP-64," Vol. 1, History & Research Division, HQ Strategic Air Command (1964).

"Recommended FY 1964 – FY 1968 Strategic Retaliatory Forces," Draft (1962).

"Recommended FY 1966 – 1970 Programs for Strategic Offensive Forces, Continental Air and Missile Defense Forces, and Civil Defense," (December 3, 1964).

"Recommended FY1968 – 1972 Strategic Offensive and Defensive Forces", Draft Memorandum for the President (1966) [theblackvault.com].

Reed, Sidney G. et al. "DARPA Technical Accomplishments" Vol. 1, IDA Paper P-2192, Institute for Defense Analysis (1990).

Reeves, James E. to L.E. Hollingsworth, U.S. AEC, Nevada Operations Office (1963).

RAND Report P-3686. Perry, Robert L. "The Ballistic Missile Decisions". Santa Monica, California: RAND Corporation (1967).

RAND Memorandum RM-3475-ARPA. Morris, Deane N. and Benson, P. "Data for ICBM Reentry Trajectories", RAND Corp. (1963).

RDA-TR-122116-001. Rosengren, Jack W. "Some Little Publicized Difficulties with a Nuclear Freeze". R&D Associates. Washington, D.C.: U.S. DoE, Office of International Security Affairs (October 1983).

"Rascal (B-63) Weapon System", Report No. 62-989-005, Bell Aircraft Corp. (Dec. 1953).

RM-1572. Kellogg, William W. and Passman, Sidney. "Infrared Techniques Applied to the Detection and Interception of Intercontinental Ballistic Missiles". Santa Monica, Calif.: Rand Corp. DTIC No. AD114 176 (October 15, 1955).

RS 3434/10. "History of the Early Thermonuclear Weapons; Mks 14, 15, 16, 17, 24 and 29". SC-M-67-661 (June 1967).

RS3434/15. "History of the Mk 30 Warhead". SC-M-67-666, Information Research Division (3434) (1967).

RS 3434/30. "History of the Mk 49 Warhead", SC-M-67-681, DoE, Information Research Division (January 1968).

RS 3434/35. "History of the Mk 54 Weapon", SC-M-67-686 (1968).

RS 3434/39. "History of the Mk 58 Warhead" SC-M-68-50 (February 1968).

RS 7331/19. Kinsey, C.H. "Re: Missile Mating Drop of a Mk-6 Re-entry Vehicle", T-18665, (1962).

SAB2001796700.  Moody, Dr. Walton S.  "History of the Joint Strategic Target Planning Staff SIOP-4 J/K, July 1971 – June 1972"

SAC200118980000.  "SAC History Development of Atomic Weapons 1956", "History Fifteenth Air Force January through June 1956" (1956).

"SAC Missile Chronology 1939 – 1988," Office of the Historian, HQ Strategic Air Command, Offutt Air Force Base, Nebraska (1990).

SAND86-2955.  Brodie, R.N.  "A Review of the U.S. Nuclear Weapon Safety Program – 1945-1986," RS3151/87/004 (1987).

SAND88-1151.  "Final Weapon Development Report for the W88 Warhead for the Mk5 Reentry Body", LANL (September 1988).  [Was classified S-RD]

SAND88-8233.  Brown, M.E.  "Development Report for the H1408 Storage & Shipping Container". (1988).

SAND91-0285.  Spletzer, Barry L.  "The Delayed Gamma Environment Produced by Exoatmosphere Nuclear Weapons Detonation".  (1992).

SAND92-0837.  Lee, Stephen R. and Barclay, Charlotte J.  "Evaluation of Factors Affecting the Timing Capabilities of the MC3858 Sprytron".  Albuquerque (1992).

SAND92-1944.  Peevy, Gregg R. et al.  "Slapper Detonator Modelling Using the PSpice Electrical Circuit Simulator".  (1993).

SAND92-2120C.  Hall, C.A. And Jacobs. E.L. "Characterization of Ceramic Capacitors for High-Voltage Pulse-Discharge Applications".  Albuquerque, NM:  Sandia National Labs (January 1, 1992).

SAND93-0943.  Seitz, T.P.  "B53-1 Special Study Report", Albuquerque, NM:  Sandia National Laboratories (June 1993).

SAND94-0335.  Hardling, D.C. et al.  "Radiant Heat Testing of the H1224A Shipping/Storage Container", Sandia Corporation (May 1994).

SAND94-0489.  Loescher, Douglas and Dinallo, Michel.  "The Response of Aeroshells to  Lightning". (1994).

SAND95-1813C.  Guidotti, Ronald A.  "Thermal Batteries:  A Technology Review and Future Directions".  CONF-951033--3 (1995).

SAND95-3065. Norwood, David P. and Martinez, Edward F.  "Evaluation of a Non-Cyanide Gold   Plating Process for Switch Tubes".  (1995).

SAND95-8004.  Brown and Higuera.  "Review of the Handling, Shipping, and Storage Equipment (H-Gear) for the W87".  (1995), handwritten notes.

SAND96-2329. Christensen, Naomi et al.  "Laboratory Directed Research and Development Final Report  Intelligent Tools for On-Machine Acceptance of Precision Machined Components" (February 1997).

SAND97-0991.  Vianco, P.T. and Rejent, J.A.  " A Microstructural Analysis of Solder Joints from the Electronic Assemblies of Dismantled Nuclear Weapons".  (1997)

SAND97-8017.  Brown, L.A. And Higuera, M.C.  "Weapon Container Catalog" Vol. 1 & 2.  (February 1998).

SAND97-8232. Hicken, W.L. et al. "Welding Development W87 Baseline" (1997).

SAND98-1459C. Archer, W.E. And Sanchez, R.O. "Manufacturing High Reliability Weapon Grade Transformers in Small Lots" (1998).

SAND98-8400. Goods, S.H. and Dombrowski, D.E. "Mechanical Properties of S-65C Grade Beryllium at Elevated Temperatures" (November 1997).

SAND99-8243. Robinson, S.L. et al. "Analysis of the Microstructure and Suitability of the First Commercial Forgings for Gas Transfer System Applications". Sandia National Laboratories (1999).

SAND2002-0307P. Loeber, Charles R. "Building the Bombs: A History of the Nuclear Weapons Complex". Albuquerque, New Mexico: Sandia National Laboratories (2002).

SAND2003-3866. Yang, Pin et al. "Chem-Prep PZT 95/5 for Neutron Generator Applications".

Sanders, M.B. and Bear, J.E. "Thermal tests of XW-47 Skirt Attachment Joint" (1961).

"Sandia Lab News", Sandia National Laboratories (26 January 2001).

"Sandstone Handbook of Nuclear Explosions", Reines, Frederick. "Scientific Director's Report of Atomic Weapon Tests At Eniwetok, 1948". (August 1, 1949).

SC-3947(TR). "Feasibility Report and Preliminary Design Proposal, Transport and Loading of New Class A, Mk 21/36, New Class B and Class C Weapons into Strike and Transport Aircraft" (December 1, 1956).

SC-DR-69-126. Pierson, H.O. and Smatana, J.F. "Development of Large Pyrolytic Carbon/Carbon Felt Cones," Sandia Laboratories, Albuquerque (June 1969).

SC-M-67-671. "History of the Mk 39 Weapon" (January 1968).

SC-M-67-680. "History of the Mk 48 Shell" (January 1968).

SC-TM-180-56-51. Compton, Joseph. "Parachute-Retardation Studies for New Class B Weapon" (1956).

SC-TM-188-56-51. McAllister, C.R. "A Comparison Between Deformation Switches and Piezoelectric Contact Fuze Devices" (October 25,1956).

SCDR 99-59. Frasier, D.D. "Test Report Jupiter Flight Test No. 13", RS 3466 (1959).

SCDR 254-59. Mecklenburg, L.W. "Corrected Neutron Generator Charge Voltages for XMC-991 Units Nos. 2001, 2003, 2005, and 2007". (January 1960).

Seaborg, Glenn T., Chairman AEC. "Office Diary, Folder-Page 024093" (December 1962).

Secretary of Defense, "Memorandum for the President: Recommended FY 1966-1970 Programs for Strategic Offensive Forces, Continental Air and Missile Defense Forces, and Civil Defense" (December 3, 1964).

Showalter, C.W. "Plutonium Blending," (June 15, 1962).

"Sigma Category Definitions", DoE HQ (October 10, 2000).

Sigmon, H. "Simulated Water Entry Shock Test of a XW-55 Type 2A Warhead". T-10633 (1964).

Siuta, T. "Final Report Fiberglass Motor Case Study Polaris Second Stage End Closure." Report No. 3642-11 (1962).

Smith, Cyril Stanley. "Minutes Discussion on Tungsten Carbide Tampers". (June 26, 1944).

Smith, Richard K. "Seventy-Five Years of Inflight Refueling Highlights, 1923–1998". Air Force History and Museums Program (1998).

Starbird, Brigadier General Alfred D. "Plumbbob Test Bulletin #43". Division of Military Application, U.S. AEC, US DOE Archives, RG: 326 U.S. Atomic Energy Commission, Collection: DMA, Box: 3780, Folder: MRA 7 Plumbob (test Bulletins – DMA) (September 4, 1957).

Starbird, Brigadier General Alfred D. "Plumbbob Test Bulletin #43". Division of Military Application, U.S. AEC, US DOE Archives, RG: 326 U.S. Atomic Energy Commission, Collection: DBM-EP Files, Box: 2702 JOB 7239, Folder: Plumbbob – Test Bulletins (September 4, 1957).

Starkey, Ryan P. et al. "Plasma Field Telemetry for Hypersonic Flight", Air Force Office of Scientific Research (2003).

Steinz, J.A. et al. "The Burning Mechanism of Ammonium Perchlorate-Based Composite Solid Propellants". Princeton University for Office of Naval Research, DTIC No. AD688 944 (1969).

"Storax III" From: A.W. Betts, "Memorandum for Chairman Seaborg [et al.]". SAC200528512001 (Dec. 17, 1962).

Strategic-Air-Command.com download of "Minuteman Missile Technology" (~1995).

T-18434. "Static Test of Filament Wound Mk-43 Center Bomb Subassembly" (1962).

"Tentative Specification for Al-B Alloy", LASL (30 November 1945).

"Test Operation Procedure 1-2-622 Vertical EMP Testing", U.S. Army Developmental Test Command (11 September 2009)]

Thorman, H. Carl. "Alleviation of Aerodynamically Induced Vibration in the Sergeant Missile", Report 20-124, Jet Propulsion Laboratory, California Institute of Technology (1959).

TM 9-3305. "Technical Manual; Principles of Artillery Weapons", HQ, Dept. of the Army, Washington, D.C. (May 1981).

T.O. 21-SM68-1. "Technical Manual, Operation and Organizational Maintenance, USAF Model HGM-25A Missile Weapon System, Operation" (1964).

Tucker, T.J. "Explosive Initiators". CONF-720303—6, Sandia Laboratories, Albuquerque (1972),

TWX RUEFHQA9007 from AFTAC, Washington, D.C. "Schooner" (December 10, 1968).

UCRL-ID-125506. Cornwell, R.S. "A Pluto-Slam Design to Fit a Maximum Fineness Ratio Missile Into a Polaris Launch Tube" (March 1962).

UCRL-ID-125523. Gibson, T.A. "Gamma Fallout Fields, Project Palanquin" (April 14, 1965).

UCRL-ID-125525, Seward, F.D. "Interim Readiness Plan" (March 7, 1969).

UCRL-ID-126119. Christie, E.R. "Neutron Prompt Burst Assembly Proposal". (October 21, 1959).

UCRL-ID-126131. "LRL Interest in Willow Testing", "Memorandum" from Steve White to Chuck Violet, (29 April 1959).

UCRL-ID-145592. Lougheed, R.W. et al. "Pu239 and 241Am (n,2n) Cross-Section Measurements Near 14 MeV" (2001).

UCRL-LR-130250. Noshkin, V.E. et al. "Radionuclides in Sediments and Seawater at Rongelap Atoll". LLNL (March 1998).

UCRL-PROC-201017. Nellis, W.J. et al. "Deuterium Hugoniot up to 120 Gpa (1.2 Mbar)" (Dec. 1, 2003).

UCRL-TR-206313. Daily, Lara D. "Simulating Afterburn with LLNL Hydrocodes" (Aug. 31, 2004).

"UK" video, "Operation Upshot/Knothole", Nevada Test Site, DoE, Accession No. 0800015 (1953; declassified in 1998).

USAF Fact Sheet 96-09. "LGM-30 Minuteman III", Department of the Air Force, downloaded off the Internet (1996).

U.S.A. vs. Wen Ho Lee, No. 99-1417-JC, U.S. District Court for the District of New Mexico. Motion Hearing, testimony of Dr. Richard Krajcik, LANL (Dec. 27. 1999).

"U.S. Strategic Objectives and Force Posture", Executive Summary, National Security Council (Top Secret) (January 3, 1971).

Wagner, Richard L. "LRL Warheads for Advanced Spartan". LRC-CMA-68-154, D00038219, LLNL (1968).

WHC-SD-CP-RPT-014. Roblyer, S.P. "Plutonium and Tritium Produced in the Hanford Site Production Reactors". (September 28, 1994).

Wainstein, L. et al. "The Evolution of U.S. Strategic Command and Control and Warning, 1945-1972", Study S-467, Arlington, Virginia: Institute for Defense Analysis (1975)

Weitze, Karen J. "Cold War Infrastructure for Strategic Air Command: The Bomber Mission," HQ Air Combat Command, Langely Air Force Base, Virginia, U.S. Army Corps of Engineers, Fort Worth District (1999).

West, A.J., SAND—95-2505. "The Evaluation of Ontario Forge Company as a Qualified Forging Vendor". Sandia National Laboratories (1981).

Weston, W.F. "Beryllium Requirements and Scrap Projections". Rockwell International. (1984).

Wheeler, Harry P., memo to Keller, L. "Re: Volume Determination of XW-40 Warhead" (October 1958).

WNP-118 "Declassification of a Material in a Specified Weapon", Classification Bulletin, Office of Classification, DoE (2008).

Wright, James B. "Appendix F to the Report of the Fundamental Classification Policy Review Group", DoE (January 17, 1997).

WX-1-E-93-410S. "Nuclear Explosive Safety Study of B53 Mechanical Disassembly Operations at the USDOE Pantex Plant", U.S. Dept. of Energy, Nuclear Explosive Safety Study Group (1993).

Wynn, Humphrey. "RAF Nuclear Deterrent Forces". London, U.K.: HMSO (1994).

WSEG [Weapons System Evaluation Group] Report No. 45. "Potential Contribution of Nike-Zeus to Defense of the Population and its Industrial Base, and the U.S. Retaliatory System." (1959).

WSEG Report No. 50. Enclosure "F", "Estimated Costs of Strategic Offensive Weapon Systems" (1960).

York, Herbert to Flaherty, J.J., letter (December 18, 1953).

Younger, Stephen, et al. "Lab-to-Lab:...Explosive-Driven Flux Compression Generators", "Los Alamos Science", Number 24 1996, Los Alamos Scientific Laboratory (1996)

## The Patent Literature


Declassified U.S. patents are mostly easy to identify by a 3 year or more difference between filing date, and issuing date. Patents are easiest to get from freepatentsonline.com on the Internet, or search google.com for several other U.S. Patent sites.

(** means special, interesting nuclear patents)

B.P. #1,605,273 (1987) ** "Missile Guidance System", Hall, Eldon C. et al., U.S. Navy; filed in 1966; classified for 23 years; Polaris A-1 or A-2 for nuclear submarine IRBM.

U.S.P. #2,588,734 (1952), "Pretreatment of Beryllium Prior to Coating", AEC.

U.S.P. #2,784,391 (1957), filed 1953, "Memory System"

U.S.P. #2,818,339 (1957), filed in 1947, "Method for Producing Malleable and Ductile Beryllium Bodies", AEC.

U.S.P. #2,844,639 (1958), nuclear battery.

U.S.P. #2,902,614 (1959), "Accelerated Plasma Source".

U.S.P. #2,913,510 (1959), nuclear battery.

U.S.P. #2,914,433 (1959), "Heat Treated U-Nb Alloys", AEC; (classified for four years).

U.S.P. #2,923,670 (1963), "Method and Means for Electrolytic Purification of Plutonium".

U.S.P. #2,937,597 (1960), "Missile Nose Structure" (filed 1956; classified 4 years).

U.S.P. #3,040,660 (1962), ** "Electric Initiator with Exploding Bridge Wire", Lawrence Johnston, (filed in 1944; classified 19 years).

U.S.P. #3,071,765 (1963; filed in 1950) "Radar Map Guidance System".

U.S.P. #3,076,408 (1963), "Controlled Fracturing of Solids by Explosives", Poulter, Thomas C. and Poncelot, Eugene F. (filed in 1955, classified 8 years).

U.S.P. #3,090,028 (1963), Electrolytic Purification of Plutonium.

U.S.P. #3,098,028 (1963), "Plutonium Electrorefining Cells".

U.S.P. #3,152, 868 (1964), "Preparation of Scandium Hydrides" (filed 1961, declassified 1964).

U.S.P. #3,158,098 (1964), Reithel, R.J. "Low Voltage Detonator System", AEC.

U.S.P. #3,170,402 (1965), "Equal Length Detonating Cords for Warhead Detonation", Morton, H.S., and Raffel, Z.M, Sec. of the Navy; (classified for 9.5 years).

U.S.P. #3,211,094 (1965), "Explosive Wave Shaper", Lidddiard, Jr., T.P., Sec. of the Navy; classified 6.5 years.

U.S.P. #3,264,147 (1966), "Beryllium Alloy and Process", Honeywell, classified 3 years.

U.S.P. #3,279,917 (1966), "High Temperature Isostatic Pressing", AEC, classified 3 years.

U.S.P. #3,281,338 (1966), "Method for Producing Ultra High Purity Plutonium Metal".

U.S.P. #3,282,806 (1966), "Electrolytic Purification of Plutonium".

U.S.P. #3,317,777 (1967), "Electrical Discharge Device", krytron, classified 5 years.

U.S.P. #3,331,744 (1967), ** "Production of Isotopes from Thermonuclear Explosions", Ted Taylor/AEC, classified for 9 years;  MICE Tritium production.

U.S.P. #3,380,687 (1968), "Satellite Spin Dispenser", Wrench, E.H. et al, General Dynamics Corp., filed 1965.

U.S.P. #3,401,019 (1968), "Process for Synthesizing Diamond", Cowan, G.R. et al.

U.S.P. #3,430,563 (1969), "Flexible Detonation Wave Shaping Device", Stresau, R.H.F., Sec. of the Navy; classified for 6.5 years;  MPI

U.S.P. #3,490,959 (1970), "Beryllium Composite", classified 4 years.

U.S.P. #3,517,615 (1970), "Explosive Wave Shaper", Jacobs, S.J., Sec. of the Navy, classified for 10 years; converging spherical implosion.

U.S.P. #3,536,544 (1970), "Trinitrotoluene Explosive Composition...", Pennington, Otis, AEC, filed in 1953, classified 18 years.

U.S.P. #3,596,604 (1971), ** "Pyrolytic Graphite Nose Tip for Hypervelocity Conical Reentry Vehicles", Air Force.

U.S.P. #3,608,490 (1971), North American Rockwell Corp., "Porous Materials", filed 1967, classified for 4 years.

U.S.P. #3,626,850 (1971), ** "Explosive Assembly", du Pont, classified for 18 years; first flyer plate patent.

U.S.P. #3,664,133 (1965), "High Acceleration Propellant Configuration".

U.S.P. #3,653,792 (1972), "High Pressure Shaped Charge Device", Garrett, D.R.; 2-point spherical implosion.

U.S.P. #3,659,423 (1972), "Moveable Rocket", Lair, Robert C. et al, Goodyear Aerospace, filed in 1964, declassified in 1972.

U.S.P. #3,659,972 (1972), "Diamond Implosion Apparatus", Garrett, D.R.; MPI levitated spherical implosion.

U.S.P. #3,682,100 (1972; classified 8 years), "Nose-Cone Cooling of Space Vehicles".

U.S.P. #3,683,190 (1972), "Tritium and Deuterium Impregnated Targets for Neutron Generators".

U.S.P. #3,716,604 (1973), ** "Method for Bonding Solid Propellants to Rocket Motor Casing";
    Hercules Inc.; filed 1967 (classified 6 years) (MM3, 3rd stage).

U.S.P. #3,724,050 (1973), "Method of Making Beryllium Shapes from Powder Metal", Berylco;
    filed 1968 (classified 4.5 years).

U.S.P. #3,743,852 (1973), ** O'Keefe, Bernard J., EG&G, "Low-Impedance, High-Voltage
    Discharge Circuit" (filed 1953; classified 20 years).

U.S.P. #3,786,448 (1974), Goodyear Aerospace Corp., "Multiple Access Plated Wire Memory".

U.S.P. #3,791,851 (1974), "Process for Heat Treating Plasma-Consolidated Beryllium", Union
    Carbide (Oak Ridge contractor) (filed 1970; classified 3.5 years).

U.S.P. #3,804,017 (1974), ** Venable et al., LANL. "Method for Mitigating Blast and Shock
    Transmission Within a Confined Volume".

U.S.P. #3,806,578 (1974), 'Purification of Pentaerythritol trinitrate by Recrystallization", Du Pont
    (filed 1960; classified 14 years)

U.S.P. #3,819,461 (1974), "Unidirectional High Modulus Knitted Fabrics", 29 patents ref. It
    (1999); classified 3 years.

U.S.P. #3,853,847 (1974), "Purification of Cyclotetramethylene Tetranitramine", Army (filed in
    1961; classified 13 years).

U.S.P. #3,869,608 (1975), "Nuclear Well Logging" (filed in 1953, classified 22 years).

U.S.P. #3,880,606 (1975), "Method of Producing a Mass Unbalanced Spherical Gyroscope Rotor",
    Rockwell International.

U.S.P. #3,880,683 (1975), "Castable High Explosive of Cyclotetramethylenetetranitramine and
    Dodecenyl Succinic Anhydride-Vinyl Cyclohexene Dioxide Polymer Binder", Navy (filed in
    1963; 11 years classified).

U.S.P. #3,883,096 (1975), "Transpiration Cooled Nose Cone", Army.

U.S.P. #3,884,735 (1975), ** Cramer, Charles H. "Explosive Composition" (filed 1956; classified for 20
    years).

U.S.P. #3,895,084 (1975), "Fiber Reinforced Composite Product", 33 patents ref. It (2000); classified 3
    years.

U.S.P. #3,894,894 (1975), "Modified Double Base Propellants with Diisocyanate Crosslinker",
    Elrick D.; Hercules (filed 1962; classified 13 years).

U.S.P. #3,896,731 (1975), "Explosive Initiator Device", Kilmer, K.E., Sec. of the Navy; classified
    for 6 years; MPI

U.S.P. #3,906,467 (1975), "Plated Wire Memory", Control Data Corporation.

U.S.P. #3,908,933 (1975), ** "Guided Missile", Goss, Willbur H. et al.; originally filed in 1956, and
    classified for 20 years; TALOS RAMjet portion of missile; an excellent 122 pages long.

U.S.P. #3,909,739 (1975), "Radiation Hardened Sense Amplifier for Thin Film Memory Applications", Air Force (filed 1969, classified 6 years).

U.S.P. #3,914,392 (1975), "High Temperature Insulating Carbonaceous Material", Sandia, DoE.

U.S.P. #3,914,395 (1975), "Process for the Manufacture of High Strength Carbon/Carbon Graphite Composites", AVCO RV's.

U.S.P. #3,922,411 (1975); AVCO; filed 1958; 17 years classified. "Honeycomb Reinforced Material".

U.S.P. #3,923,496 (1975), DoE, (filed 1944, classified 31 years). "Nickel Powder..."

U.S.P. #3,925,122 (1975), ** "Molded Explosive Bodies Having Variable Detonation Speeds", Dynamite Nobel; classified for 6 years; plastic microspheres to vary HE detonation velocity.

U.S.P. #3,956,039 (1976), ** "High Explosive Compound", Crawford/LASL, classified for 21 years; Boracitol, binary HE lens.

U.S.P. #3,956,658 (1976), ** "Low Impedance Switch", Donald Hornig, LASL (filed in 1945; classified for 32 years).

U.S.P. #3,976,888 (1976), ** "Fission Fragment Driven Neutron Source", Miller, Lowell G. et al. U.S. ERDA.

U.S.P. #3,985,595 (1976), High-Purity TATB.

U.S.P. #3.993,738 (1976), "High Strength Graphite and Method for Preparing Same", Oak Ridge, DoE; classified 3 years.

U.S.P. #3,997,899 (1976), "Low Radar Cross-Section Re-Entry Vehicle", Rolsma; classified for 16 years.

U.S.P. #4,008,411 (1977), ** "Production of 14 MeV Neutrons by Heavy Ions", Brugger, Robert M. et al., U.S. ERDA.

U.S.P. #4,014,979 (1977), "Method of Producing Wurtzite-Like Boron Nitride", Dremin et al.

U.S.P. #4,041,872 (1977), ** "Wrapper, Structural Shielding Device", Army (filed 1971; classified 6 years); Spartan ABM missile shielding.

U.S.P. #4,056,662 (1977), "Thermal Cell and Electrolyte Composition Therefor" (filed in 1954, classified 23 years).

U.S.P. #4,089,267 (1978), "High Fragmentation Munition", Mescall et al.

U.S.P. #4,098,625 (1978), HMX PBX with Teflon-Viton binder; Navy (filed in 1968, classified 10 years).

U.S.P. #4,100,322 (1978), "Fiber-Resin-Carbon Composites and Method of Fabrication", McDonnell Douglas Corp.; 19 patents ref. It (1999); classified 3.5 years.

U.S.P. #4,112,179 (1978) "Method of Coating With Ablative Heat Shield Materials". 20 patents ref. It (2000).

U.S.P. #4,123,597 (1978), "Thermal Cells" (filed in 1953, classified 25 years).

U.S.P. #4,147,108 (1979), "Warhead"; AAI Corp., filed in 1955, classified for 25 years; HE/flyer plate gap.

U.S.P. #4,147,822 (1979), "Composite Structure and Process and Apparatus for Making the Same", McDonnell Douglas, filed in 1976.

U.S.P. #4,152,381 (1979), ** "Method for Preparing Metallated Filament-Wound Structures", Oak Ridge, DoE; hardened RV.

U.S.P. #4,163,231 (1979; filed in 1968) Raytheon, "Radar Mapping Technique".

U.S.P. #4,185,558 (1980), "Re-entry Vehicle Boundary Layer Transition Suppressor", US Air Force (filed in 1968; classified for 12 years).

U.S.P. #4,187,782 (1980), "Shaped Charge Device", Army.

U.S.P. #4,326,117 (1982), ** "Weld Braze Technique"; boosting gas bottle.

U.S.P. #4,334,474 (1982), ** "Warhead Initiation System", Coltharp, David R., Navy (filed 1976; classified 6 years); MPI.

U.S.P. #4,350,297 (1982), "Swivelling Exhaust Nozzles for Rocket Motors" (filed 1962, classified 20 years).

U.S.P. #4,359,732 (1982), "Topographical Mapping Radar"; Goodyear Aerospace (filed 1963; classified 19 years.

U.S.P. #4,370,576 (1983), ** "Electric Generator"; Foster, John, LLNL (filed in 1962; classified 21 years; first FCG).

U.S.P. #4,430,132 (1984), "Desensitizing Explosives", U.K. Government (filed in 1980; classified 4 years).

U.S.P. #4,482,129 (1984), "All Metal Valve Structure for Gas Systems". DoE, Miamisburg, Ohio.

U.S.P. #4,482,405 (1984), "Explosive Molding Composition and Method for Preparation Thereof", Holtston Ordnance, (filed in 1960, classified for 24 years).

U.S.P. #4,490,329 (1984), Hare et al. "Implosive Consolidation…"

U.S.P. #4,502,096 (1985), ** Reynolds Industries, Inc.; "Low-Inductance Capacitor".

U.S.P. #4,517,497 (1985), ** Reynolds Industries, Inc.; "Capacitor Discharge Apparatus".

U.S.P. #4,552,742 (1985), ** "Materials Processing Using Chemically Driven Spherically Symmetrical Implosions", Mayer, F., KMS Fusion.

U.S.P. #4,577,812 (1986), "Centrifugally Operated Moving-Mass Roll Control System", Platus, Daniel H.; Air Force (filed 1973; classified for 12 years); W-68 RV.

U.S.P. #4,602,565 (1986), "Exploding Foil Detonator", Reynolds Industries (filed 1983), slapper.

U.S.P. #4,608,222 (1986), ** "Method of Achieving the Controlled Release of Thermonuclear Energy", Brueckner, KMS Fusion (filed in 1973, classified 13 years), ICF.

U.S.P. #4,628,819 (1986), "Disintegrating Tamper Mass", Dept. of the Navy.

U.S.P. #4,673,430 (1987), ** International Nickel Company, Toronto, Canada (filed 1947, classified 41 years); U. hex gaseous diffusion barrier.

U.S.P. #4,711,086 (1987), "Trident II First and Second Stage Internal Insulation"; Navy.

U.S.P. #4,788,913 (1988), ** Stroud, John. "Flying-Plate Detonator Using A High-Density High Explosive", (filed 1971, declassified after 17 years).

U.S.P. #4,790,735 (1988), ** "Materials Processing Using Chemically Driven Spherically Symmetrical Implosions", Mayer, F., KMS Fusion; MPI.

U.S.P. #4,835,304 (1989), "High Density Ester Damping Fluids", Charles Stark Draper Lab.

U.S.P. #4,860,655 (1989), "Implosion Shaped Charge Perforators", Chawla, M.S. (filed 1985).

U.S.P. #4,952,255 (1990), "Extrudable PBX Molding Powder", Navy (filed in 1984; classified 6.5 years).

U.S.P. #4,982,665 (1991), "Shaped Charge"; Navy, filed in 1973, classified for 17 years; HE/flyer plate gap.

U.S.P. #4,498,367 (1985), "Energy Transfer Through a Multi-Layer Liner for Shaped Charges".

U.S.P. #4,993,662 (1991, filed 1970) Navy, "Apparatus for Guiding a Missile".

U.S.P. #5,004,185 (1991, filed 1964, classified 27 years), "Air-Surface-Missile Data Link System", Navy.

U.S.P. #5,059,839 (1991), "Explosive Magnetic Field Compression Generator Transformer..."; Navy (filed 1977; classified 14 years).

U.S.P #5,070,789 (1991), "Electrical Exploding Bridge Wire Initiators".

U.S.P. #5,074,937 (1991), "Preparing an Elastomeric Bound Explosive", Naval Weapons Center, China Lake, CA (filed 1975; classified 16.5 years).

U.S.P. #5,090,324 (1992), "Warhead", Rheinmetall, Germany.

U.S.P. #5,132,080 (1992), Inco., filed 1944, classified 48 years. "Production of articles from powdered metals".

U.S.P. #5,301,612 (1994), ** "Carbon-Assisted Flyer Plates", LASL.

U.S.P. #5,322,020 (1994), "Shaped Charge", Bernard et al., classified for 11 years.

U.S.P. #5,332,560 (1994), "High Density-High Purity Graphite Prepared by Hot Isostatic Pressing in Refractory Metal Containers", DoE/LLNL, classified for 4 years.

U.S.P. #5,336,520 (1994), "High Density-High Purity Graphite Prepared by Hot Isostatic Pressing in Refractory Metal Containers", DoE/LLNL, classified 4 years.

U.S.P. #5,359,935 (1994), "Detonator Device and Method for Making Same", Willett, J.D.; MPI.

U.S.P. #5,444,598 (1995), "Capacitor Exploding Foil Initiator Device", Arescu, Carmelo.

U.S.P. #5,450,794 (1995), "Method for Improving the Explosive Performance of Underwater Explosive Warheads", Drimmer, Bernard; classified for 33 years.

U.S.P. #5,479,860 (1996), "Shaped-Charge with Simultaneous Multi-Point Initiation of Explosives", Ellis, J.

U.S.P. #5,549,731 (1996), "Preparation of Solid Aggregates of High Density Boron Nitride Crystals", Cline, Carl F.. et al.

U.S.P. #5,565,644 (1996), ** "Shaped Charge With Wave Shaping Lens", MPI.

U.S.P. #5,689,084 (1997), (filed in 1974, classified 24 years!). "Bonding Method…"

U.S.P. #5,600,088 (1997), Oberth, A. "Coating for Solid Propellants"; Aerojet-General. (filed in 1988; classified 9 years).

U.S.P. #5,717,397 (1998), "Low Observable Shape Conversion for Aircraft Weaponry", Lockheed.

U.S.P. #5,908,802 (1999), Voigt, J.A. et al. "Nonaqueous Solution Synthesis Process for Preparing Oxide Powders of Lead Zirconate Titanate and Related Materials".

U.S.P. #6,174,493 (2001), "Porous Beryllium"; Oak Ridge, DoE (filed in 1967; classified 34 years).

U.S.P. #6,662,546 (2003), "Gas Turbine Engine Fan", GE (classified 10 years).

U.S.P. #6,760,396 (2004), "Coated Metal Articles and Method of Making", DoE (classified 58 years!) .

U.S.P. #6,761,862 (2004), Oak Ridge, (filed 1945, classified 59 years), UF6 barrier integrity testing.